CW01188785

Studies in Renaissance Literature

Volume 39

THE ATOM IN SEVENTEENTH-CENTURY POETRY

Studies in Renaissance Literature

ISSN 1465-6310

General Editors
Raphael Lyne
Sean Keilen
Matthew Woodcock
Jane Grogan

Studies in Renaissance Literature offers investigations of topics in English literature focussed in the sixteenth and seventeenth centuries; its scope extends from early Tudor writing, including works reflecting medieval concerns, to the Restoration period. Studies exploring the interplay between the literature of the English Renaissance and its cultural history are particularly welcomed.

Proposals or queries should be sent in the first instance to the editors, or to the publisher, at the addresses given below; all submissions receive prompt and informed consideration.

Dr Raphael Lyne, Murray Edwards College, Cambridge, CB3 0DF

Professor Sean Keilen, Literature Department, UC Santa Cruz, 1156 High St, Santa Cruz, CA 95064, USA

Professor Matthew Woodcock, School of Literature and Creative Writing, University of East Anglia, Norwich, NR4 7TJ

Dr Jane Grogan, University College Dublin, School of English, Drama and Film, Newman Building, Belfield, Dublin 4

Boydell & Brewer Limited, PO Box 9, Woodbridge, Suffolk, IP12 3DF

Previously published volumes in this series are listed at the back of this volume

THE ATOM IN
SEVENTEENTH-CENTURY POETRY

Cassandra Gorman

D. S. BREWER

© Cassandra Gorman 2021

All Rights Reserved. Except as permitted under current legislation no part of this work may be photocopied, stored in a retrieval system, published, performed in public, adapted, broadcast, transmitted, recorded or reproduced in any form or by any means, without the prior permission of the copyright owner

The right of Cassandra Gorman to be identified as the author of this work has been asserted in accordance with sections 77 and 78 of the Copyright, Designs and Patents Act 1988

First published 2021
D. S. Brewer, Cambridge

ISBN 978-1-84384-593-5

D. S. Brewer is an imprint of Boydell & Brewer Ltd
PO Box 9, Woodbridge, Suffolk IP12 3DF, UK
and of Boydell & Brewer Inc.
668 Mt Hope Avenue, Rochester, NY 14620–2731, USA
website: www.boydellandbrewer.com

A catalogue record for this title is available
from the British Library

The publisher has no responsibility for the continued existence or accuracy of URLs for external or third-party internet websites referred to in this book, and does not guarantee that any content on such websites is, or will remain, accurate or appropriate

This publication is printed on acid-free paper

Printed and bound in Great Britain by
TJ Books Ltd, Padstow, Cornwall

For Ted

CONTENTS

Acknowledgements ix
Conventions and Abbreviations xi

Introduction 1

1 Atomic Congruity: The Philosophical Poetry of Henry More 37
2 Thomas Traherne's Atoms, Souls and Poems 75
3 World-Making and World-Breaking: The Atom Poems of Margaret Cavendish and Hester Pulter 117
4 The Atom in Genesis: Lucy Hutchinson's *Order and Disorder* 175

Afterword: A Poetics of the Atom 215

Bibliography 223
Index 241

ACKNOWLEDGEMENTS

I have many people to thank for their support while I worked on this book. Many thanks go to Boydell & Brewer and to my editor, Caroline Palmer, for her generous help. An anonymous reader gave productive feedback on an initial draft and outline. I am deeply grateful, moreover, to Liza Blake, for her exceptionally detailed, careful and constructive feedback on the entire manuscript. Her comments have been invaluable to completing this project.

From my time at the University of Cambridge, where my fascination with literary atoms began, my great thanks go to Joe Moshenska, David Hillman and Jason Scott-Warren for their indefatigable encouragement and advice. Kathryn Murphy, John Rogers, Claire Preston, Stephen Clucas, Antony Ossa-Richardson, James Jiang, Alexander Wragge-Morley, Ned Allen, Austen Saunders and Alexander Freer also provided support and exceedingly helpful feedback at the early stages of conceptualising and writing this book. At Anglia Ruskin University, I am grateful to the Faculty of Arts, Humanities and Social Sciences for the support of a research sabbatical while I finished this project. Special mention should go to Lizzie Ludlow and Tory Young, who have been enormously supportive. Further thanks go to Sarah Brown and Emma Pauncefort, who both kindly read and offered feedback on draft chapters. In the summer of 2018 I was fortunate to receive a fellowship at the Folger Shakespeare Library, which lent me all-important protected time for thinking and drafting. I extend my great thanks to the librarians there as well as to those at the Cambridge University Library, Beinecke

Acknowledgements

Rare Book and Manuscript Library, British Library and Bodleian Library for their assistance. I am grateful to Boydell & Brewer for their permission to reprint in Chapter Two some passages from my essay 'Thomas Traherne and "Feeling Inside the Atom"', in *Thomas Traherne and Seventeenth-Century Thought*, ed. Elizabeth S. Dodd and Cassandra Gorman (Cambridge, 2015), pp. 69–83. Thanks go moreover to the journals *Literature Compass* and *The Seventeenth Century* for their permission to reproduce small extracts from my articles 'Poetry and Atomism in the Civil War and the Restoration' (2016) and 'Lucy Hutchinson, Lucretius and Soteriological Materialism' (2013), in the Introduction and Chapter Four.

The generous scholarly community of *Scientiae: Disciplines of Knowing in the Early Modern World* listened to me give papers on Hester Pulter and Henry More at two annual conferences, where I received excellent questions and suggestions. I am grateful to Subha Mukherji, Elizabeth Swann and all those involved with the *Crossroads of Knowledge* project at the University of Cambridge, for invitations to present on early modern literature and natural philosophy and for many inspiring conversations. Thanks go moreover to the Thomas Traherne Association, for the invitation to speak at the Thomas Traherne Festival some years ago and for the helpful feedback I received there. Conversations with Beth Dodd, with whom I co-organised a symposium on Traherne back in 2012, also greatly enriched my work on the Herefordian theologian and poet.

Closer to home, thanks go to Elena Field, Yamna Khan and Dilyana Mihaylova for their kindness and excellence as living companions while I was completing the book. I owe great thanks to my parents, Gail and Robert Gorman, for their unswerving love and support through every stage of this process and beyond. Final thanks go to Ted Morcaldi, who has brought love, joy, music and negronis into my life. This book is dedicated to him, with much love.

CONVENTIONS AND ABBREVIATIONS

All quotations from the Bible, unless specified otherwise, are taken from *The Bible: Authorized King James Version with Apocrypha*, ed. Robert Carroll and Stephen Prickett (Oxford, 1997). Quotations from the Geneva Version in Chapter Four are taken from *1599 Geneva Version: The Holy Scriptures Contained in the Old and New Testaments*, ed. Peter A. Lillback and Marshall Foster (Dallas, 2006). All transcriptions of manuscript material are semi-diplomatic; I have maintained original spelling, punctuation and capitalisation but expanded abbreviations (and italicised the supplied text).

ABBREVIATIONS

edn	edition
ELH	*English Literary History* (journal)
fol./fols	folio(s)
GNV	Geneva Version of the Bible
MS/MSS	manuscript(s)
ODNB	*Oxford Dictionary of National Biography*, ed. H. C. G. Matthew and Brian Harrison, 60 vols (Oxford, 2004); online edition at <http.//www.oxforddnb.com/>
OED	*Oxford English Dictionary*; online edition at <http://www.oed.com/>
PMLA	*Publications of the Modern Language Association of America*
sig. / sigs	signature(s)

INTRODUCTION

> They say then that Love was the most ancient of all the gods; the most ancient therefore of all things whatever, except Chaos, which is said to have been coeval with him... And himself out of Chaos begot all things, the gods included. The attributes which are assigned to him are in number four: he is always an infant; he is blind; he is naked; he is an archer... The fable relates to the cradle and infancy of nature, and pierces deep. This Love I understand to be the appetite or instinct of primal matter; or to speak more plainly, *the natural motion of the atom*; which is indeed the original and unique force that constitutes and fashions all things out of matter. Now this is entirely without parent; that is, without cause... [T]he summary law of nature, that impulse of desire impressed by God upon the primary particles of matter which makes them come together, and which by repetition and multiplication produces all the variety of nature, is a thing which mortal thought may glance at, but can hardly take in.
>
> (Francis Bacon, 'Cupid, or the Atom', *De Sapientia Veterum*)[1]

In his 1609 publication *De Sapientia Veterum*, Bacon sets out to discover the hidden, allegorical truths embedded in thirty-one figures from classical mythology. Though elsewhere he urges that 'science must eliminate the fable and focus on facts', his Preface to

---

[1] *The Wisdom of the Ancients* (English translation of *De Sapientia Veterum*), in *The Works of Francis Bacon*, ed. James Spedding, Robert Leslie Ellis and Douglas Denon Heath, 14 vols (Boston, 1857–74), XIII, pp. 67–172 (pp. 122–23).

De Sapientia Veterum argues for the continued necessity of parables and similitudes to advance learning.[2] These not only sweeten the pill when introducing new theories – allowing the philosopher to communicate without 'offence or harshness' – but also sustain communion with the divine in the pursuit of knowledge, evoking the very use of typological 'veils and shadows' that 'religion delights in'.[3] Allegories, even as they cloak and protect their hidden significance, allow the learner a glimpse of purer, more distant truths.

Memorably, Bacon includes amongst his mythologies the myth of Cupid, or 'Love', whom he equates with the 'appetite or instinct of primal matter; or to speak more plainly, *the natural motion of the atom*'. In the section 'Cupid, or the Atom', Bacon traces the cryptic history of a complex figure who bears two forms from his classical origins. The newest and most recognisable of these is the cherubic infant, 'youngest of all the gods, son of Venus'. According to other strands of thought, however, Cupid – or Eros, as he appears in Hesiod – owes his birth to primordial beginnings, as the 'most ancient of all the gods' alongside the original Chaos.[4] It is this Cupid, Bacon argues, who assumes the form of an atom. He expounds some key parallels between the god of Love and the indivisible particle to support his claim. Atoms, like Cupid's arrows, dart and pierce; like Cupid, atoms were considered 'blind', as they jostle, collide and congregate in space; like the infant god, atoms are simultaneously ancient and yet fresh from birth – the first and last of all things.

---

[2] Quoted from Liza Blake and Kathryn Vomero Santos, 'Introduction', in *Arthur Golding's 'A Moral Fabletalk' and Other Renaissance Fable Translations*, ed. Liza Blake and Kathryn Vomero Santos (Cambridge, 2017), pp. 1–47 (p. 7). Bacon famously dismisses fable from the remit of his new learning, as outlined in the *New Organon*. For more information, see Blake and Vomero Santos, pp. 7–8.

[3] Preface to *The Wisdom of the Ancients*, pp. 76–80.

[4] See Hesiod, *Theogony*, 116: 'In truth, first of all Chasm came to be, and then broad-breasted Earth ... and Eros, who is the most beautiful among the immortal gods, the limb-melter – he overpowers the mind and the thoughtful counsel of all the gods and of all human beings in their breasts.' Quoted from *Hesiod: Theogony, Works and Days, Testimonia*, ed. and trans. Glenn W. Most (Cambridge, MA, 2006), p. 13.

Appropriately, given Bacon's overall interest in the original, pregnant 'mystery' of ancient fables, his account of 'Cupid, or the Atom' develops some interesting complexities. His opinions on atomism altered throughout his philosophical career, but his fascination with '*the atom*' had little to do with scientific truth.[5] Atoms, according to Bacon, were of great use as an instrument or theory for further discovery in natural philosophy, regardless of whether they held actual status in the structure of material things. They deepened enquiry and provided a useful hypothesis for considering physical properties. His writing on the atom in *De Sapientia Veterum* is nevertheless largely unconcerned with scientific practice. Instead, as Silvia Manzo has explained, Bacon presents us with an '*ontological* account of the atom' (italics mine).[6] His focus is on the original impulse behind matter, how motion and form came to be, for which the necessary cognitive tools are feeling and intuition rather than understanding. Mortal thought 'may glance' at these atomic beginnings, but 'can hardly take in' – this realisation heightens the mission of allegory, which is especially multi-layered and dense in this fable. The 'Cupid' of common understanding, the cherub of Eros armed with bow and arrow, himself is a typological representation of the primordial archetype behind creation, 'the most ancient of all the gods'. It is then at a deeper level still that these two Cupids come to represent the original particle of material life.

Bacon's metaphysical and metaphorical account of 'Cupid, or the Atom' anticipates the many ways in which the literary atom amassed imaginative power and potential over the coming century. The early modern atom, most simply and singularly, was an indivisible particle. On the one hand, its inaccessibility to thought and representation made it unfathomable; on the other, its constant, enduring presence from the beginning of things to their end suggested it held the key to ontological enquiries (related to the origins and nature of being.) Bacon is careful not to leave out God

---

[5] Silvia A. Manzo discusses this further in 'Francis Bacon and Atomism: A Reappraisal', in *Late Medieval and Early Modern Corpuscular Matter Theories*, ed. Christoph Lüthy, John E. Murdoch and William R. Newman (Leiden, 2001), pp. 209–43 (p. 221).

[6] Ibid., p. 221.

in his reflection on first principles, confirming that it was the deity who 'impressed' the 'impulse of desire' upon the atoms. As Manzo has neatly analysed, the atom is 'a factor of continuity ... persisting throughout the shift from Chaos to cosmos. Through the divine Word, atoms go from disorder to order, producing orderly atomic structures without losing their essential primary form of matter.'[7] Manzo's words correspond with Bacon's, who implies that meditation on 'Cupid, or the Atom' can lead us back to the original divine impression. The atom emerges as a connecting point between immortality and the mortal, material experience. Her analysis could apply equally to the poetics of the twenty-book creation narrative *Order and Disorder* (1679).[8] Lucy Hutchinson, its author, was certainly not an advocate of atomism in scientific practice, but she found comfort in the parallels between atomic reconfiguration and the regenerative properties of the divine Word. Bacon's use of allegory to explore first principles also anticipates the Cambridge Platonist Henry More's *Philosophicall Poems* (1647), which combined Neoplatonic allegory with an atomic focus that 'pierces deep' (to borrow Bacon's phrase) into the origins of creation.[9] Moreover, the sentiment of casting a 'glance' at this ancient origin resonates with Thomas Traherne's writings on the atom from the 1670s, which connected the indivisible particle with the infinite enquiring powers of the immortal soul. As with Bacon's essay, these fascinations with the atom are not works of experimental natural philosophy nor, for the most part, critical responses to the details of Epicurean atomism.[10] They

---

[7] Ibid.
[8] The date provided refers to the poem's incomplete publication (of the opening five books), *Order and Disorder; or, The World Made and Undone. Being Meditations upon the Creation and the Fall* (London, 1679). The full manuscript is Yale, Beinecke Library, Osborn MS fb 100. David Norbrook attributed the poem to Lucy Hutchinson in 'Lucy Hutchinson and *Order and Disorder*: The Manuscript Evidence', *English Manuscript Studies 1100-1700*, 9 (2000), 257-91. He then published a complete edition: *Lucy Hutchinson: Order and Disorder*, ed. David Norbrook (Oxford, 2001).
[9] Henry More, *Philosophicall poems* (Cambridge, 1647).
[10] Thomas Traherne poses the slight exception, as he writes 'Of Epicurisme' in his commonplace book and bases his atomic theory in part on the

Introduction

reveal instead a rich poetics stemming from different influences, enmeshed in metaphysical and ontological concerns. This is not a book about the literary reception of philosophical atomism in early modern England. Rather, I reveal that many poets and philosophers of the period sought positive, spiritual motivation in the concept of the indivisible particle, in discourses distanced from the threats of Epicurean materialism and atomism. The poets of this study recognised an affinity between atoms and poems, wherein both could form points of contraction, or blank canvases for the gathering of ideas. Poetry provided a developmental space for associations of the atom with spiritual endurance and divine comfort. From these positive, creative forays into indivisibility, a poetics of the atom surfaces in seventeenth-century literature. This poetics, rather than mourning the dissolution of coherence into 'atomies', moves in the opposite direction, finding strength in the indivisibility of the particle and in its power to direct the imagination to things that 'pierce deep.'[11]

Previous works of literary and historical scholarship have introduced an early modern cultural reception of atomism that resisted its doctrines: Christian theologians and philosophers associating atomic theory with atheism and hedonism; poets in mourning for the breaking of the Platonic 'circle'; and the reduction of all material beings into insentient machines.[12] These fears and criticisms

---

accounts of Epicurean philosophy by Diogenes Laertius and Theophilus Gale. See his commonplace book, Bodleian, MS Eng. poet. c. 42 (known as the 'Dobell Folio'), 41v–42v; I discuss the influence of Diogenes Laertius and Gale upon Traherne in Chapter Two of the current book, pp. 83–84.

[11] A well-known example of destructive 'atomies' is found in John Donne's *Anniversaries*: 'And freely men confess that this world's spent, / When in the planets, and the firmament / They seek so many new; they see that this / Is crumbled out again to his atomies.' 'An Anatomy of the World', in *The Poems of John Donne, Vol. 1: The Text of the Poems with Appendixes*, ed. Herbert J. C. Grierson (Oxford, 2012), pp. 231–45 (p. 237).

[12] See, for example, Marjorie Hope Nicolson, *The Breaking of the Circle: Studies in the Effect of the 'New Science' upon Seventeenth Century Poetry* (Evanston, 1950) and Stephen Clucas, 'Poetic Atomism in Seventeenth-century England: Henry More, Thomas Traherne and "Scientific Imagination"', *Renaissance Studies*, 5.3 (1991), 327–40. For classic critical

5

were current in seventeenth-century literature and this book will not be ignoring them. Most of these negative responses, however, were directed at the ultra-materialism of Epicurean philosophy, which inevitably created metaphysical and theological tensions. Assumptions that every reference to an 'atom' would belong to philosophical atomism has withheld recognition of the particle's broader, extended poetic life.

In recent years, scholarship on atomism in early modern culture has responded to this neglect with a decisive shift towards poetics, though this has arisen from renewed attention on the most materialist of poets: Lucretius, the author of *De Rerum Natura* – a long, didactic poem on the principles of Epicureanism. The major discovery of this research is that the early modern atomic revolution was more specifically a revolution in philosophical poetry. Copies of Lucretius's *De Rerum Natura* spread fast across early modern Europe. In her study *Reading Lucretius in the Renaissance*, Ada Palmer declares that 'Fifty-four surviving manuscripts and thirty print editions of the poem were produced between 1417 and 1600.'[13] Palmer distinguishes in her work between those who read Lucretius and those who were atomists. She takes as her main interest

> how Lucretius's atomism was rescued from the dust by men who were not atomists, how his radical heterodoxies were copied and preserved by men who were not radicals, and how his scientific theories were injected back into scientific discourse by poets and philologists rather than scientists.[14]

This taps into one of the most important factors of Lucretius's influence: that the growing interest in the beauty of his poetry – 'which the great Virgil himself had deigned to imitate' – introduced a

---

works on the mechanisation of early modern natural philosophy, see A. Rupert Hall, *The Scientific Revolution, 1500–1800: The Formation of the Modern Scientific Attitude* (Boston, 1954), E. J. Dijksterhuis, *The Mechanization of the World Picture: Pythagoras to Newton*, trans. C. Dikshoorn (Princeton, 1986) and Herbert Butterfield, *The Origins of Modern Science, 1300–1800*, 2nd edn (New York, 1965).

[13] Ada Palmer, *Reading Lucretius in the Renaissance* (Cambridge, MA, 2014), p. 4.

[14] Ibid., p. xiii.

multitude of readers to Epicurean atomism and was entwined with the growing popularity of atomic thought in early modern natural philosophy.[15] The Renaissance preoccupation with Lucretius's aesthetic hints, moreover, at how an atomic poetics could manifest and disperse throughout early modern writings. Just three years prior to Palmer's monograph, Gerard Passannante's *The Lucretian Renaissance* broke new ground with its central thesis that 'the history of materialism in the Renaissance' was 'inextricably tied to a history of literature and the material text'.[16] Passannante's argument is prompted especially by one of Lucretius's most pervasive analogies, the similitude between atoms and letters of the alphabet, and the question of how 'the idea that letters are like atoms' became 'meaningful for the development of textual criticism'.[17] Some writers and readers took Lucretius's 'vision of flux and change' as sign of 'an increasingly crumbling and fragmented tradition on the brink of resolving into chaos'; others, Passannante acknowledges, found 'a figure of creative potential, progress, and generativity'.[18] While Passannante's topic is Lucretius, his methodology shares important sympathies with my own. My study similarly excavates the presence of an early modern atomic poetics that inspired 'creative potential, progress, and generativity', though not in this case through the tropes of Lucretian atomism. In seventeenth-century English poetry, the utmost indivisibility of the

---

[15] Ibid., p. 4.
[16] Gerard Passannante, *The Lucretian Renaissance: Philology and the Afterlife of Tradition* (Chicago, 2011), p. 4.
[17] Ibid. Lucy Hutchinson translates a typical passage featuring Lucretius's atom-and-letters analogy:
    For the same principles doe constitute
    Heave'n, earth, sea, sun, floods, creatures, plants, & fruite.
    But divers ways their divers beings compound.
    So in our verse are comon letters found
    In severall transpositions sett, from whence
    Words are producd, of severall sounds and sence.
The Works of Lucy Hutchinson, Vol. 1: The Translation of Lucretius, ed. Reid Barbour, David Norbrook and Maria Zerbino (Oxford, 2012), Book 1, lines 832–37.
[18] Passannante, *The Lucretian Renaissance*, p. 8.

invisible atom broke free from its materialist bonds, assuming new meaning from its power as a tool for contemplating the nature of self, society, natural world and God.

The aim of this book is to move beyond the popular paths of scholarship on *De Rerum Natura*, towards recognising and tracing a poetics of the atom that was inspired by the concept of indivisibility rather than the features of Lucretian atomism. My work is indebted nonetheless to the approaches of recent studies on the reception of Lucretius, which have been ground-breaking in relocating poetry at the centre of early modern discourses of materialist science. It is necessary, therefore, to introduce their findings in some detail here. Passannante's work has taken the lead in a scholarly movement that has re-examined the Renaissance reception of Epicurean materialism, exploring the influence of Lucretius through the lens of poetics rather than through the reiteration of philosophical doctrines.[19] He claims that 'the history of materialism as a philosophy is also fundamentally a history of words and poetic tropes'.[20] This focus forms the background for Jessie Hock's 2021 monograph *The Erotics of Materialism: Lucretius and Early Modern Poetics*, which argues 'seventeenth-century poets read [*De Rerum Natura*] as a treatise on the poetic imagination, initiating an atomist genealogy at the heart of the lyric tradition'.[21] Hock's compelling study encourages a rethinking of early modern poetic and specifically lyric forms, acknowledging a previously understated Lucretian influence on their shape and content: '[b]ecause Lucretius expounds his poetics in a language of love and desire, the influence of Lucretius in early modernity is particularly potent

---

[19] In addition to the pre-quoted works by Passannante and Palmer, studies include Jacques Lezra, *Unspeakable Subjects: The Genealogy of the Event in Early Modern Europe* (Stanford, 1997); Jonathan Goldberg, *The Seeds of Things: Theorizing Sexuality and Materiality in Renaissance Representations* (New York, 2009); Adam Rzepka, 'Discourse *Ex Nihilo*: Epicurus and Lucretius in Sixteenth-Century England', in *Dynamic Reading: Studies in the Reception of Epicureanism*, ed. Brooke Holmes and W. H. Shearin (Oxford, 2012), pp. 113–32; and Jessie Hock, *The Erotics of Materialism: Lucretius and Early Modern Poetics* (Philadelphia, 2021).

[20] Passannante, *The Lucretian Renaissance*, p. 5.

[21] Hock, *The Erotics of Materialism*, p. 3.

in love poetry'.[22] Her thesis suggests it was Lucretius's *poetic* atoms, rather than explicitly the atomism of Epicurean philosophy, that had the greatest impact on early modern literary culture. A focus on Lucretian poetics nevertheless presents challenges for the critic in addition to its extensive opportunities. In an essay on the English literary reception of Lucretius, Adam Rzepka re-emphasises the claim that '*De Rerum Natura* provides a model for the poetics of its own subsequent transmission', arguing that this took place in the sixteenth century as 'the surface effect of a more complex, manifold circulation of tropes that had not yet cohered as public discourse'.[23] The challenge, as Rzepka admits, is in how to define and identify these tropes when they are frequently unsignposted, and in how to reconcile them with the adamant protestations against Lucretius and Epicurean philosophy elsewhere in early modern culture. A recent collection of essays, *Lucretius and the Early Modern*, further discusses the tensions and contradictions of the Lucretian reception. In the introduction, David Norbrook acknowledges the issues at hand, declaring the focus of the collection on the 'ways in which Lucretian themes and imagery percolated early modern culture, in the face of often subtle resistances' and 'the constraints as well as the possibilities offered by early modern conditions of textual production and circulation'.[24]

It is worth noting that these tensions, far from putting early moderns off their Lucretius, seem to have generated increasing scholarly interest. By the sixteenth century, Lucretius was on the syllabus at the universities and the most renowned material philosophers of the seventeenth century – including Pierre Gassendi, René Descartes, Thomas Hobbes and Robert Boyle – were influenced deeply by atomic ideas.[25] Over the course of the following century *De Rerum Natura* was widely translated into English, though only

---

[22] Ibid., p. 8.
[23] Rzepka, pp. 116 and 122.
[24] 'Introduction', in *Lucretius and the Early Modern*, ed. David Norbrook, Stephen Harrison and Philip Hardie (Oxford, 2016), pp. 1–27 (p. 12).
[25] See Mordechai Feingold, 'The Humanities', in *The History of the University of Oxford, Volume IV: Seventeenth-Century Oxford*, ed. Nicholas Tyacke (Oxford, 1997), pp. 211–359 (p. 250).

a few known complete translations survive: Lucy Hutchinson's full version from the 1650s; an anonymous prose translation housed in the Bodleian library; and Thomas Creech's celebrated English edition from 1682.[26] The diarist and Royal Society fellow John Evelyn translated at length from the poem, publishing the first book with an attached commentary on Gassendi in 1656; in the Restoration, John Dryden and John Wilmot, the 'Libertine' Earl of Rochester, also translated popular passages.[27] A large part of the poem's appeal lay in its controversy. Lucretius addresses his poem to a friend, Memmius, whom he is instructing on Epicurean physics, metaphysics and morals. He propounds the atomic philosophy as an antidote to fears and superstitions bred by organised religion – there is nothing to fear, Lucretius argues, because all things are material, even the soul; as a consequence, all natural forms are necessarily finite in a world of constant movement and change. *De Rerum Natura* introduces Epicurus's theories of atoms and void, and the ways in which atoms congregate to produce all natural things. In six books Lucretius expounds the Epicurean understanding of generation and corruption, motion, the senses, the nature of the soul, creation and the history of mankind and meteorology, before concluding incompletely and infamously with an explicit account of plague in Athens. The poem was commonly denounced as dangerously irreligious, and references to Lucretius or Epicurus frequently accompanied early modern insults.[28] Some of these deserve to be taken with a

---

[26] The manuscript copy of Lucy Hutchinson's translation is British Library, Add. MS 1933; the anonymous prose translation is Bodleian Library MS Rawl. D. 314; Creech's published version is *Lucretius his six books of Epicurean philosophy and Manilius his five books containing a system of the ancient astronomy and astrology together with The philosophy of the Stoicks*, trans. Thomas Creech (Oxford, 1682).

[27] Evelyn's translation is now in the British Library, Evelyn Collection MS 34 and MS 34a. The first book was published as John Evelyn, *An essay on the first book of T. Lucretius Carus De Rerum Natura. Interpreted and made English verse* (London, 1656). Dryden's translations of selected passages were published in Creech's *Lucretius his six books* (1682). For Rochester, see 'Translations from Lucretius', in *The Works of John Wilmot Earl of Rochester*, ed. Harold Love (Oxford, 1999), pp. 108–09.

[28] See Clucas, 'Poetic Atomism', 334.

pinch of salt, as the pull of Lucretius's poetry more than occasionally outweighed the trepidation of his atheistic values. Lucy Hutchinson prefaced her translation by expressing utter disdain for Lucretius, calling him 'Lunatick', 'Dog' and a 'crabbed poet', but she nevertheless remained committed to her full study of his poem, offering a presentation copy to Arthur Annesley, the Earl of Anglesey, in 1675.[29]

It would be too simplistic, however, to conclude that poetic mastery overcame philosophical controversy in the early modern reception of Lucretius. As David West argued in his seminal work *The Imagery and Poetry of Lucretius*, 'the philosophical subject matter of this poem is not an impediment to the poetry, it is rather the stimulus for the impassioned observation and contentious contemplation of the material world which contribute so much to the poetic intensity of the work'.[30] Philosophy and poetry, West argues, are permanently entwined in a passionate and intense contemplation of nature. This level of 'poetic intensity' has been a crucial point of interest for readers and critics of Lucretius. Responding to his previous work on Lucretian Epicureanism and early modern culture, Gerard Passannante has recently considered the relationship between philosophical materialism, 'impassioned observation' and 'poetic intensity' further in his book *Catastrophizing: Materialism and the Making of Disaster*. He identifies a theme of 'catastrophic materialism' across early modern texts, which he links to Epicureanism and defines as 'a reasoning from the sensible to the insensible, a precipitous shift or collapse of scale and perspective, a temporal compression of beginning and end, and an act of imaginative making'.[31] His is not a study of atomism per se, but Passannante raises the prevalent Lucretian simile of motes in the sunbeam as an example of a materialist image that can fill the mind, resulting in experiences of 'catastrophic materialism' that were recorded in early modern texts.[32]

---

[29] See 'To the Right Honorable Arthur Earle of Anglesey', in *The Works of Lucy Hutchinson*, I, pp. 5–15 (p. 7; p. 13).
[30] David West, *The Imagery and Poetry of Lucretius* (Edinburgh, 1969), p. 17.
[31] Gerard Passannante, *Catastrophizing: Materialism and the Making of Disaster* (Chicago, 2019), p. 4.
[32] Ibid., p. 102. ('The notion of infinity turns back again to the analogy that begins small and suddenly occupies the imagination.')

Lucretian materialism may have filled minds, but there was greater significance to the atoms of early modern literature. What the studies of Passannante, Palmer and Hock have done for Lucretian poetics I aim to achieve here by introducing an alternative, yet equally pervasive, poetics of the atom in seventeenth-century culture. By this poetics, I understand the following. Firstly, and most simply, that it was through its exploration in poetry that the atom accrued greater significance in early modern culture. Secondly, that seventeenth-century poets recognised an affinity between atomic movement and the construction of poetic forms: in a way parallel to the particles that continuously combine and recombine to form all things, poetry that contemplates atoms enacts a self-conscious awareness of its own power to build worlds, knowledge and identities. Thirdly, and most importantly, that there were sympathies between the concept of the single, indivisible atom and the creative capacity of the poetic space, which materialised (to use an apt word) in the attempts of poetic forms to depict and explore the otherwise unfathomable entity. Atoms and poems share an important characteristic: inherently, they bear multiplicity of meaning even amidst seeming simplicity; as they take form, they exercise their potential to accommodate diverse possibilities, complexities and paradoxes.[33] Within the texts discussed in this book, poems, as well as atoms, emerge as means for guiding the imagination on its search for answers to profound ontological questions. From this perspective, an alternative poetics of the atom comes into focus, distinct from the Lucretian reception, with the spiritual endeavour to reconnect with the *metaphysical* nature of things. Over the following chapters, Henry More merges atomic thought with a

---

[33] It is also important that the atom, an entity that cannot be observed, requires multiple explanations to be conceived. Jessie Hock has noted this in an article on Lucretius and Margaret Cavendish's *Poems and Fancies*: 'Certainty could be achieved by closer observation, which is unfortunately impossible; with no access to a more privileged perspective on the true causes of things, humans have to make do with "probabillitie." Multiple explanations, then, offer a menu of probable causes from which the reader can choose.' See Hock, 'Fanciful Poetics and Skeptical Epistemology in Margaret Cavendish's *Poems and Fancies*', *Studies in Philology*, 115.4 (2018), 766–802 (778).

Neoplatonic aesthetic to discover the power of self-reflection; Thomas Traherne develops a poetics that realises the productivity of the atom by way of preserving its active relationship with a superior indivisible, the soul; Margaret Cavendish and Hester Pulter realise the empowerment of selfhood promised by atomic acts of world-breaking and world-making; and Lucy Hutchinson rewrites Lucretian poetics with the combined influences of the Geneva Bible and corpuscular alchemy.

Within literary studies, the most prominent critical works on the alternate, literary lives of the early modern 'atom' remain writings by Stephen Clucas and Reid Barbour from the close of the twentieth century. Clucas's article 'Poetic Atomism in Seventeenth-Century England' sets out to consider how 'the atom haunted the imagination of its detractors and its supporters' during the period.[34] 'Haunted' is a key word within this study as, deftly and persuasively, Clucas proceeds to demonstrate some of the many ways in which the atom gathered negative associations in seventeenth-century literature. Beginning with the crumbling of Donne's 'atomies' in the *Anniversaries*, he moves from cosmic disintegration to political allegories and fears of 'autokinetic atoms'. He concludes, however, with a brief but illuminating nod to the particles found in the writings of Henry More and Thomas Traherne, which served a positive spiritual purpose. Clucas finishes by acknowledging that '[t]he corollary of these hysterical narratives of disintegration are narratives of reintegration and unison in the mind of God'.[35] Seven years later, Barbour's book-length study *English Epicures and Stoics* teases out several of these observations in greater detail, but the divine significance of the atom does not fall within its remit.

---

[34] Clucas, 'Poetic Atomism', 328. Clucas has published additional studies of early modern atomism. See 'The Atomism of the Cavendish Circle: A Reappraisal', *The Seventeenth Century*, 9.2 (1994), 247–73; 'Corpuscular Matter Theory in the Northumberland Circle', in *Late Medieval and Early Modern Corpuscular Matter Theories*, pp. 181–208; and '"All the mistery of infinites": Mathematics and the Atomism of Thomas Harriot', in *Mathématiques et connaissance du monde réel avant Galilée*, ed. S. Rommevaux (Montreuil, 2010), pp. 113–54.

[35] Clucas, 'Poetic Atomism', 335 and 340.

Barbour's thesis is 'that early Stuart culture is diacritically obsessed with the Stoics and Epicureans'. He reveals, in chapters on the reception of Epicureanism and Stoicism respectively, how 'early Stuart readers' applied ideas and tropes from the philosophical schools in response to 'many of their most urgent political and religious questions'.[36] The chapters on Epicureanism introduce an array of literary 'atoms' in writings of the period, from the prevalent atom/fairy metaphor to 'analogies between the atom and the sovereign', notably in the Jacobean and Caroline court masque. From political contexts Barbour moves to consider the resonance of Epicurean ideas in theology, though his focus is more concerned with the paradoxical lending of Epicurean ethics to meditations on 'fundamental religious matters' than it is with the significance of the atom.[37]

These important studies laid the foundations for future work on the imaginative possibilities of seventeenth-century atoms and atomisms. My focus responds to the research of Clucas and Barbour, but the argument of this book reveals another side to the early modern literary atom: that the atom accumulated a poetics of its own in writings that embraced its indivisibility and related it closely to human experience. Clucas's article is revelatory in its introduction to the poetic richness of the indivisible particle, but his topic is its relations to theories of atomism, not 'the atom' as it could be separated from materialism. His closing claim, that atoms also encouraged 'narratives of re-integration and union' with God, introduces the other side to 'narratives of disintegration' but is not the focal point of his study.[38] Barbour's subject is the reception of Epicureanism, which is not my chief concern in the current work. My focus, moreover, is the mid and late seventeenth century, where Barbour's study concludes with the argument that, by the 1660s, 'Dissenters ... tended to renounce whatever Pauline accommodations they had once made with Seneca, Epicurus, and the pantheon of pagan philosophers.'[39] Barbour suggests that the

---

[36] Reid Barbour, *English Epicures and Stoics: Ancient Legacies in Early Stuart Culture* (Amherst, 1998), p. 2.
[37] Ibid., p. 7; p. 78.
[38] Clucas, 'Poetic Atomism', 340.
[39] Barbour, *English Epicures*, p. 244.

failed Interregnum may have led Republicans and Independents to turn their back on a past interest in Epicurean values and morals: these were theories that could be associated more readily, come 1660, with the restored orthodoxy. This did not, however, break dissenting ties with the poetic power of the indivisible atom. On the contrary, the poetics of the atom beyond Lucretius and Epicureanism grew stronger over the course of the Restoration as the commonality of the term, along with its expanded alternate meanings, spread across literary contexts.[40]

A focus on the atom may seem minute, but it resonates richly across a broader context of intellectual thought. The significant place of imaginative literature in early modern natural philosophical enquiry has been well-considered in recent criticism. Studies such as those by Elizabeth Spiller and Frédérique Aït-Touati have shown how the forms of seventeenth-century fiction were productive in the process of making knowledge: Spiller by dissolving the lines between fact and fiction in early modern textual knowledge-making; Aït-Touati by tracing a 'shared imagination' between the natural philosophical and the poetic, with an especial focus on the 'poetics of the cosmological discourse of the period'.[41] It is now a general understanding that the lines between fiction and natural philosophy were blurred in early modern literary culture, and that they were mutually informing as knowledge-seeking texts. This is also Claire Preston's thesis in *The Poetics of Scientific Investigation in Seventeenth Century England*, which argues that, while the influence of the 'empirical behaviours' of early modern science upon the contemporary literature is obvious ('if inadequately recognized'), 'less obvious and more contentious is the claim that scientific expression, the behaviour of natural philosophers, and even the shape and structure of their investigations have a literary

---

[40] For an exploration of this, see Cassandra Gorman, 'Poetry and Atomism in the Civil War and Restoration', *Literature Compass*, 13.9 (2016), 560–71.

[41] Elizabeth Spiller, *Science, Reading, and Renaissance Literature: The Art of Making Knowledge, 1580–1670* (Cambridge, 2004); Frédérique Aït-Touati, *Fictions of the Cosmos: Science and Literature in the Seventeenth Century*, trans. Susan Emanuel (Chicago, 2011), p. 4.

and rhetorical basis'.[42] The current book responds to the aims of these studies but with a different focus. My subject is not primarily the 'poetics of scientific investigation', but how poetry reimagined and recontextualised the concept of the atom *beyond* the natural philosophical sphere. Renaissance poetics typically praised the singularity of poetry as the most privileged art, more capable of engaging with higher things than other modes of expression.[43] The sympathetic relationship between atom and poetry was mutually enforcing: the subject of the indivisible prompted poets to push at the very limits of what could be grasped, towards the unfathomable; an emerging poetics of the atom, meanwhile, extended the life of the particle well beyond scientific engagements with atomism and materialism.

'Poetics' is a broad term, but it rests generally on the study of 'creative principles' – a category immediately relevant to both the techniques of poetry and the building blocks of atoms.[44] I will

[42] Claire Preston, *The Poetics of Scientific Investigation in Seventeenth-Century England* (Oxford, 2015), p. 10.

[43] As Philip Sidney writes famously in *The Defence of Poesy*, 'Nature never set forth the earth in so rich tapestry as divers poets have done ... her world is brazen, the poets only deliver a golden.' Quoted from *Sidney's 'The Defence of Poesy' and Selected Renaissance Literary Criticism*, ed. Gavin Alexander (London, 2004), pp. 1–54 (p. 9). George Puttenham moreover opens *The Art of English Poesy* with the claim that 'the name and profession' of 'poet' has 'no small dignity and pre-eminence, above all artificers scientific or mechanical ... this science in his [the poet's] perfection cannot grow but by some divine instinct – the Platonics call it *furor* – or by excellency of nature and complexion'. See *Sidney's 'The Defence of Poesy' and Selected Renaissance Literary Criticism*, pp. 55–203 (pp. 57–58). Francis Bacon also comments on the special liberty of poetry, albeit with a word of caution, in *The Advancement of Learning*, where he writes that poetry 'may at pleasure ioyne that which Nature hath seuered: & seuer that which Nature hath ioyned'. Quoted from *The Oxford Francis Bacon IV: The Advancement of Learning*, ed. Michael Kiernan (Oxford, 2000), p. 73. Bacon's words echo Philip Sidney's praise of the poet who 'doth grow in effect another nature, in making things ... better than Nature bringeth forth'. Anna-Maria Hartmann makes this connection in *English Mythography in its European Context, 1500–1650* (Oxford, 2018), p. 142.

[44] I cite 'creative principles' from the *OED*, 'Poetics, n. 1.b'.

Introduction

trace a suggestive intimacy between poetic and atomic forms in writing of the period, a recurrent and evolving bond between the atom and the poetic space in literary acts of ordering, dismantling and creating. Imaginative and spiritual responses to the indivisible particle not only permeated seventeenth-century poetry but also influenced the shape and movements of poetic forms. The poets and philosophers of this book – Henry More, Thomas Traherne, Margaret Cavendish, Hester Pulter, Lucy Hutchinson and John Milton – all in their various ways tapped into the atom to explore, as did Francis Bacon, that which could be felt but not directly experienced or expressed. Empirically speaking, an atom was incomprehensible; for many, unable to conceive of the actuality or consequences of indivisibles in matter, the theory was unthinkable from a rationalist perspective, too. Yet for those who did accept the *possibility* of the atomic hypothesis, the smallest units of the material world – closest to creation and closest to God – redirected reasoning and offered, metaphysically and intuitively, a unique opportunity to approach the deepest secrets of the world and the divine.

It is in relation to this remarkable positive, metaphysical and spiritual side to the atom in the seventeenth century that poetry arises as a dominant form. Some things, said Traherne, 'are if not Known by Causes, Felt by Intuition': it was with poetry that he chose to conclude his section on atomic theory, and to make his most contentious arguments.[45] For the writers in this book, the poetic medium offered an experimental space, increased freedom of expression, and a popular outlet for the development of ideas. It was in such a vein that Margaret Cavendish opened her first collection of literary writings with a series of 'atom poems', each lyric of which experiments with an aspect of the atomic concept – and in so doing builds a fabric to preserve the author's fame. Moreover, as explicitly stated by Henry More but endorsed by each writer, poetry came close to achieving the status of a Platonic ideal, a song motivated by and in contact with higher principles.[46] It is no coin-

---

[45] Thomas Traherne, *Commentaries of Heaven*, in *The Works of Thomas Traherne*, ed. Jan Ross, 9 vols (Cambridge, 2005–), III (2007), p. 341.
[46] Henry More, for example, writes in the preface to his *Opera Philosophica* that he was 'stir'd' to compose his first poem, *Psychozoia*, by 'some

cidence that the conceptions of the atom outlined in this book were directed through this vehicle, the form best suited to expressing the essence of an entity that was otherwise unknowable – and, through its endurance and indivisibility, a common emblem for resurrection and even the immortal soul.

The authors discussed in this study employed a variety of poetic forms to embrace the atom. For More, it was philosophical poetry by way of Spenser, as the philosopher adopted the stanza form of *The Faerie Queene* to build didactic, and often allegorical reflections on the nature of the cosmos.[47] Traherne's interest in the atom owed much to his reading of More's work, but his use of poetic form did not: Traherne wrote devotional lyrics after George Herbert as his means to consider the indivisible.[48] Lyric was moreover the form of choice for both Margaret Cavendish and Hester Pulter's 'atom poems', though the latter also wrote 'emblems' for experimental reflections

---

Heavenly Impulse of Mind'. More, Preface to the *Opera Philosophica*, trans. Richard Ward. Quoted from Richard Ward, *The Life of Henry More by Richard Ward*, ed. Sarah Hutton, Cecil Courtney, Michelle Courtney, Robert Crocker and Rupert Hall (Dordrecht, 2000), p. 21. I discuss this further in Chapter One, p. 63. Margaret Cavendish presents a slight exception to this focus on divinity. Her privileging of poetic fiction nevertheless still adopts a super-natural (in the literal sense of the compound term) approach to what the powers of creative thought can achieve; namely, the regeneration and immortality of authorial self, an endeavour set in motion by her early 'atom poems'. This poetic act is technically based in materiality as, according to Cavendish's philosophy, all things – even disembodied thoughts – are material. I discuss the complexities of this philosophy further in Chapter Three, pp. 122–23.

[47] More's collections of poetry are *Psychodia Platonica; or, a Platonicall Song of the Soul* (Cambridge, 1642); *Democritus Platonissans, or, An essay upon the infinity of worlds out of Platonick principles* (Cambridge, 1646); and *Philosophicall poems* (Cambridge, 1647).

[48] I discuss the poems at the end of Traherne's section on 'The Atom', from the *Commentaries of Heaven*, in Chapter Two. I also analyse lyrics from two manuscripts of Traherne's poetry: the 'Dobell Folio', now Bodleian, MS Eng. poet. c. 42, and the 'Poems of Felicity', now British Library, Burney MS 392.

Introduction

on moral subjects.[49] Lucy Hutchinson's *Order and Disorder* shares the epic form of *Paradise Lost* even while rejecting its ambitions, justifying its status as a biblical creation poem without the classical influence.[50] The forms of poetic expression differ, but there is a key similarity binding each author's work into a recognisable poetics of the atom. Each atomically influenced text is concerned with orders of sequence and growth, while it contracts its creative focus on to the singularity of the ultimate particle. The particles of More's stanzas accumulate into a poetic whole that provides the foundation for his burgeoning philosophy. Traherne's lyrics 'follow' individual atoms on a journey inwards that, as his poetry progresses, develops into an increased understanding of self, world and the divine.[51] Cavendish and Pulter's lyrics are presented as sequential studies; like the atom, each poem stands alone but also connects with the verses surrounding it, mimicking and enabling the personal 'world making' that emerges as a shared, dominant concern. As declared on her frontispiece, Hutchinson's very focus in her creation poem is the nature of 'Order and Disorder'. Her narrative crushes together experiences of physical disorder with the promise of spiritual (and atomic) restoration,

---

[49] Cavendish's 'atom poems' open her first publication, *Poems, and fancies* (London, 1653). Hester Pulter compiled her original lyrics and emblems into a manuscript that is now Leeds, Brotherton Collection, MS Lt q 32. I discuss these works further in Chapter Three.

[50] While Milton famously aspires, at the opening of *Paradise Lost*, to soar 'with no middle flight' and to 'assert the eternal providence, / And justify the ways of God to men', Hutchinson writes in *Order and Disorder* as though she were aware of Milton's invocation and disapproves of its ambition. She requests, more cautiously:
Give utterance and music to my voice,
Singing the works by which thou art revealed.
What dark Eternity hath kept concealed
From mortals' apprehensions, and what been
Before the race of time did first begin,
It were presumptuous folly to inquire.
John Milton, *Paradise Lost*, ed. Alistair Fowler (Harlow, 2007), Book 1, 14–26; *Lucy Hutchinson: Order and Disorder*, Canto 1, 36–41.

[51] On the significance of 'following' an atom in Traherne's work, see Chapter Two, pp. 102–03.

19

supported throughout by references to the biblical passages that permit an extraordinary absorption of past, present and future forms of being. These works, like the atoms that form their subjects and their textures, are concentrated on acts of ordering and disordering; dissolving and creating. The poems of these authors are exquisitely self-conscious atomic forms, but they are not merely aware of their place within the corpuscular world. To return to my earlier point about the metaphysical purpose of the poetic atom – the literary indivisible that assumes power surpassing its roles in natural philosophy – the atoms of these poems direct their texts beyond the mundane, producing intimate reflections on the creative power and indivisibility of self, soul and God.

On this note, before moving forward with the intentions of this study, it is necessary to clarify a few distinctions. As I have asserted, there is quite a gulf between the discourses surrounding philosophical 'atomism' and those inspired by the 'atom', singular, in the period. Works of sixteenth- and seventeenth-century lexicography typically define an atom as, rather broadly, an extremely small thing. To give just two examples: Thomas Blount refers to the atom as 'any thing so small that it cannot be made less' in his *Glossographia* (1661); Edward Phillips, nephew of John Milton, defines the subject as 'a mote in the sun-beams, also a word used in Philosophy, being the smallest part of a body that can be imagined' in *The new world of English words* (1658).[52] Such emphasis on the atom as the smallest thing conceivable introduces a term with great imaginative potential, equipped to accommodate a great many ideas. Its near synonymity with the mote, moreover, was inspired by an analogy from *De Rerum Natura* but soon outgrew its Lucretian origins to generate diverse meanings of the 'atom' in early modern culture: atoms became dust, fairies, insects and other

---

[52] Thomas Blount, *Glossographia; or, A dictionary interpreting all such hard words of whatsoever language now used in our refined English tongue* (London, 1661), sigs E6r–v. Edward Phillips, *The new world of English words* (London, 1658), sig. Dv.

Introduction

minute bodies.[53] In contrast, atomism took many forms as a philosophical matter theory during the period, as I will introduce in the following section.

Additionally, there is often a sharp distinction in the literature between the atom and its plural, 'atoms' or 'atomies', which when depicted in motion frequently accompany images of dissolution, threat and chaos. This book responds to these divergences by exploring the poetics of the atom in what can be considered two halves. In the chapters on Henry More and Thomas Traherne I focus largely on the significance of the 'atom' singular, which represents stability in contrast to the atomic swarm, and indivisible permanence in opposition to physical corruption. The second half of the book is more concerned with 'atoms' plural, but not in relation to

---

[53] The key passage for the atoms-as-motes analogy comes from *De Rerum Natura*, Book 2. Lucy Hutchinson translates:
    For if you marke, when the high sun conveys
    Into an obscure roome his piercing rayes,
    Even where the light flowes in with glorious streames,
    Armies of attoms sport in those bright beames (2. 112–15)
The full passage is 2. 108–25. See also Reid Barbour's discussion of the 'mythologies of atomism,' p. 48 (and pp. 44–48 for a discussion of the fairy analogy). There are fascinating early modern parallels between atoms and small creatures, notably insects. In his *Theatrum Insectorum* (supposedly complete by 1589; first published, posthumously, in 1634), Thomas Muffet comments on 'small lice' or 'hand-worms', arguing 'our eye can scarsely discern them, they are so small, that *Epicurus* said it was not made of Atoms, but was an Atom it self'. The idea of insects, or minute creatures, as 'living atoms' took off in the seventeenth century, to the extent that, by 1664, the clergyman Ralph Brownrig could describe the swarm of flies during the Old Testament plague with the following: 'Who can order, or direct the tumultuary motions of living Atoms, and errattick Creatures?' In the same year, Henry Power – experimental philosopher and atomist – referred to a 'Whey-worm' as a 'living Atom'. Thomas Muffet, *Theatrum Insectorum*, in Edward Topsell, *The history of four-footed beasts and serpents describing at large their true and lively figure, their several names, conditions, kinds, virtues ... countries of their breed, their love and hatred to mankind, and the wonderful work of God in their creation, preservation and destruction ... whereunto is now added, The theater of insects, or, Lesser living creatures* (London, 1658), p. 1095; Ralph Brownrig, *Twenty five sermons* (London, 1664), p. 222; Henry Power, *Experimental Philosophy* (London, 1664), p. 22.

the destructive movements of atomisation. Margaret Cavendish, Hester Pulter and Lucy Hutchinson all, in their various ways, wrote poetry that meditated on the liberating power of atoms to dissolve and recongregate into renewed and resurrected forms. This poetic history of the irreducible particle has not been written by historians of science; neither has it been studied by those who have identified a poetics of atomism via the influence of Lucretius. In the following section of this introduction I will outline, very briefly, a history of atoms and atomism in seventeenth-century philosophy, the philosophical and intellectual background against which a poetics of the atom evolves but takes a novel direction.

ATOMS AND ATOMISM

Classically, atomism is a philosophy that argues that the material world is composed of minute, indivisible particles (atoms) which move at random in empty space (void).[54] As individual atoms in motion meet and collide, they congregate together to create recognisable material forms. There have been many variants on the philosophy, not least in seventeenth-century Europe, but there is little more to its essence than this. It is largely to the Hellenic philosophers that early modern western Europe owed its atomic renaissance.[55] The theory that all things came into being through the movements of atoms in void was asserted first by Leucippus and

---

[54] For definitions of classical atomism, see *The Hellenistic Philosophers*, ed. and trans. A. A. Long and D. Sedley, 2 vols (Cambridge, 1987), *Volume 1: Translations of the Principal Sources with Critical Commentary*, p. 6; Andrew Pyle, *Atomism and its Critics: From Democritus to Newton* (Bristol, 1997), p. 19 (on 'The Democritean Beginnings').

[55] It must be acknowledged, however, that this statement offers far too simple a history. Corpuscular theories continued to be developed during the Middle Ages, but were not directly indebted to the Pre-Socratic philosophers. The atomism of Democritus and Epicurus was almost absent from early medieval Europe (with the exception of brief acknowledgements in the work of early encyclopaedists) but continued to be developed by Islamic philosophers, who may in turn have been influenced by theories of Indian origin. For more information on this see Pyle, pp. 210–12. See also Shlomo Pines, *Studies in Islamic Atomism*, trans. Michael Schwarz and ed.

Introduction

his pupil Democritus in the fifth century BC, then expanded upon substantially a few generations later by the post-Socratic philosopher Epicurus.[56] It has previously been claimed that between late antiquity and the fifteenth century, which saw the rediscovery of *De Rerum Natura* and the publication of Diogenes Laertius's *Lives of the Eminent Philosophers*, atomism lay dormant and undeveloped, a theoretical proposition that was unknown to the scholastic philosophers of medieval Europe.[57] This was not entirely the case.

From the twelfth century onwards, the possibility of atomism had begun to grow in European interest. The major influence for this – at this stage, minor – resurgence was Aristotle and the dominating scholasticism of the universities. Aristotle was no atomist; he argued against the theories of the pre-Socratic atomists (including Anaxagoras, whom he considered to be of the clique) in his *Physics*, *De Caelo*, *Metaphysics* and *De Generatione et Corruptione*.[58] It was nevertheless through these critiques that the

---

Tzvi Langermann (Jerusalem, 1997), pp. 117–41, on the relations between Indian and Kalam atomism.

[56] For further information on this, see Long and Sedley, p. 504.

[57] Pyle, for example, claims that 'During the Dark Ages of Western Christendom the Atomic Theory – like so much else – was almost totally forgotten ... There is no trace of a continuous and developing tradition of Atomist thought' (pp. 210–11). Marie Boas Hall opened her influential article 'The Establishment of the Mechanical Philosophy', *Osiris*, 10 (1952), 412–42, with the claim: 'The application of atomic theory to physical science began, as is well known, in the seventeenth century' (413). Stephen Greenblatt also claims early Christianity 'set the stage for the subsequent disappearance of the whole Epicurean school of thought' in *The Swerve: How the Renaissance Began* (London, 2011), p. 98.

[58] Pyle discusses these references from Aristotle in *Atomism and its Critics*, pp. 14 and 21–22. He then proceeds to analyse Aristotle's 'extended critique' of indivisibles in the *Physics*, pp. 25–30. Aristotle also refers to the atomism of his opponents in *De Anima*. See *De Anima (On the Soul)*, trans. Hugh Lawson-Tancred (London, 1986), 403b31–404a5. Theophilus Gale writes of Epicurus in *The Court of the Gentiles*: '*Laertius* affirmes, he was chiefly addicted to *Anaxagoras*'. Theophilus Gale, *The court of the gentiles: or A discourse touching the original of human literature, both philologie and philosophie, from the Scriptures, and Jewish church*, 3 vols, p. 440. Laertius writes that 'Among all the Ancient Philosophers, he approv'd Anaxagoras'.

details of atomic philosophy reached a wide scholastic readership. Denouncements and occasional explanations of atomism in the works of Augustine, Lactantius and other Church Fathers accompanied the 'more balanced and analytical discussions', as Catherine Wilson has put it, of Cicero and Plutarch.[59] A number of European philosophers proposed corpuscular matter theories in advance of the early modern Epicurean renaissance, from the widely known and influential (Roger Bacon; Ramon Lull) to the more obscure (the Oxonian fourteenth-century philosophers, Henry of Harclay and Walter Chatton).[60]

Moreover, in denouncing Democritean theories of matter Aristotle suggested his own particulate alternative, what became known by his commentators and followers as the doctrine of 'natural minima'. The following argument from the *Physics* was greatly responsible for developing these ideas:

> If ... it is impossible for an animal or a plant to be of any size whatsoever, in the direction of greatness and of smallness, it is evident that none of its parts can be either [of any size whatsoever] ... [I]f every limited body is done away with by [subtracting] a limited body:

See *The lives, opinions, and remarkable sayings of the most famous ancient philosophers written in Greek, by Diogenes Laertius; to which are added, The lives of several other philosophers, written by Eunapius of Sardis; made English by several hands* (London, 1696), Book X, p. 191.

[59] Catherine Wilson, *Epicureanism at the Origins of Modernity* (Oxford, 2008), p. 14.

[60] Christoph Lüthy, John E. Murdoch and William R. Newman discuss the fourteenth-century atomists Henry of Harclay and Walter Chatton in their 'Introduction' to *Late Medieval and Early Modern Corpuscular Matter Theories*, pp. 1–38 (pp. 8–12). Earlier still, three prominent European figures of the thirteenth century demonstrate the range of backgrounds possible in medieval atomism: Roger Bacon, who made a significant contribution to the understanding of Aristotelian *minima*; his older contemporary Robert Grosseteste, who developed an atomism of 'Democritean, Platonist and Pythagorean extraction' (ibid., p. 20); and Ramon Lull, whose atomic theory 'of the continuous and discrete powers of things' was to inspire the later Neoplatonic and hermetic atomism of Nicholas of Cusa and Giordano Bruno. On Lull, I quote from Charles Lohr, 'Ramon Lull's Theory of the Continuous and Discrete', in *Late Medieval and Early Modern Corpuscular Matter Theories*, pp. 75–90 (p. 88).

> it is evident that it is not possible for everything to be present in everything. For if flesh is subtracted from water, and if this is done again by segregation from what remains, even if what is subtracted is always smaller, still it will not be smaller than a certain magnitude.[61]

The complexities of this argument, both in its original context and in terms of its long reception, are too weighty a topic for the current study. It is enough just to draw one important conclusion: Aristotle firstly suggests the possibility that minimally small parts of a body may exist, then bases his argument on the diminishing of flesh on the assumption that they do. In his comprehensive study *Atomism and its Critics*, Andrew Pyle voices words of caution at this point to calm the over-excited reader – Aristotle's 'natural minima' only surface in the context of his anti-Anaxagoran critique, and are spoken of only as potential parts; they are not applied to non-living matter; they have no appearance in the philosopher's *De Generatione et Corruptione*.[62] The theory also contradicts one of Aristotle's most defining natural philosophical arguments, the conceptualisation of the material world as an infinitely divisible continuum. It was his commentators, and not the Stagirite himself, who offered solutions to this problem. As Pyle explains:

> In the works of Alexander, Philoponus and Simplicius the doctrine receives more definitive formulation. *Qua* mathematical extension, they say, quantity is potentially infinitely divisible; physically, it is not. Each type of substance has its *natural minimum*, beyond which it cannot be further divided[.][63]

The principle of the *minima* remained potential, rather than an assertion of actuality. In Pyle's words, 'One could not trace the life-history of a natural minimum, as one could of an Epicurean atom.'[64]

Yet it is to some extent possible to trace a history of the minimum through the scholastic Middle Ages and Renaissance, whether it was physically actual or not. While, as Andrew G. van Melsen admits, 'Aristotle's theory of smallest particles' was still in

---

[61] Aristotle, *Physics*, 1.4, 187b1–188a1. Quoted from Aristotle, *Physics*, trans. C. D. C. Reeve (Indianapolis, 2018), pp. 8–9.
[62] Pyle, p. 216.
[63] Ibid.
[64] Ibid., p. 217.

its 'embryonic stage', it reached maturity in the scholastic philosophy of Averroes, Aquinas and, considerably later, Joseph Scaliger and Daniel Sennert.[65] The latter two were notable in applying the doctrine of natural *minima* to practical chemistry. Throughout the early modern period, corpuscular theories and alchemical practice were closely connected. The presence of *minima* (and, for some, indivisibles) in matter explained various phenomena from chemical experimentation, for philosophers from pseudo-Geber to the later figureheads of the 'new science', Robert Boyle and Isaac Newton.[66]

It is therefore not as clear-cut to argue, as did many of the traditional historians of the scientific revolution, that atomism arose in the seventeenth century as a reaction against scholasticism.[67] Recent studies in the history of science have been firm in

---

[65] Andrew G. van Melson, *From Atomos to Atom: The History of the Concept Atom* (New York, 1960) p. 43.

[66] Many works have been published on the corpuscular philosophies of Robert Boyle and Isaac Newton. See especially William R. Newman, *Newton the Alchemist: Science, Enigma, and the Quest for Nature's 'Secret Fire'* (Princeton, 2019); William R. Newman, 'Boyle's Debt to Corpuscular Alchemy' in *Robert Boyle Reconsidered*, ed. Michael Hunter (Cambridge, 1994), pp. 107–18; Antonio Clericuzio, *Elements, Principles and Corpuscles: A Study of Atomism and Chemistry in the Seventeenth Century* (Dordrecht, 2000), pp. 103–48 (on Boyle); William R. Newman, *Atoms and Alchemy: Chymistry and the Experimental Origins of the Scientific Revolution* (Chicago, 2006), pp. 157–215 (on Boyle). Newman introduces Pseudo-Geber and discusses the philosopher's influence upon Daniel Sennert in *Atoms and Alchemy*, pp. 26–34.

[67] As Dennis Des Chene writes: 'In the historiography of seventeenth-century natural philosophy, it was long customary to divide philosophers into two camps: the *scolastici*, or those who continued to adhere to the Aristotelianism that had predominated in the schools of Europe for several centuries; and the *novatores*, those who like Descartes, Boyle, or Gassendi, rejected fundamental principles of Aristotelianism'. 'Wine and Water: Honoré Fabri on Mixtures', in *Late Medieval and Early Modern Corpuscular Matter Theories*, pp. 363–79 (p. 363). See also Marie Boas Hall, 'The Establishment of the Mechanical Philosophy': 'From the mid-sixteenth century there was a new interest in the underlying nature of matter and the application of atomic theories to the new science just freeing itself from Aristotelian doctrines. This movement was initially a part of the general anti-Aristotelian attitude engendered by the Renaissance humanists' (p. 423).

reassessing the role of Aristotelianism in corpuscular matter theories, most notably the work of William R. Newman and Antonio Clericuzio.[68] With this in mind, however, it is still undeniable that the atomic renaissance of the seventeenth century was indebted to the resurfacing of Epicureanism. The year 1473 saw the first Latin printing of the classical biographer Diogenes Laertius's *The Lives, Opinions, and Remarkable Sayings of the most famous Ancient Philosophers*, the tenth and final book of which focuses entirely on Epicurus. Laertius introduced Epicurus in a very sympathetic light, attempting to undo the negative character sketches presented by succeeding classical philosophers and the Fathers. He argues that there are 'sufficient Testimonies of this Man's undeniable and his exceeding Candor and Civility towards all Persons'; that he had so many friends 'that whole Cities were not able to contain 'em'; and that 'his inclinations of Piety toward the Gods ... were beyond Expression'.[69] This portrayal of the atomist created new opportunities for Christian philosophy to engage with Epicureanism. For several influential natural philosophers of early modern Europe, including a number of corpuscular matter theorists, this Christian

---

Andrew G. Van Melsen also claims 'philosophic atomism criticized the very essence of Aristotelian philosophy' in *From Atomos to Atom*, though he does devote a chapter to corpuscular theories within scholasticism (p. 58).

[68] Historical scholarship of recent years, notably work by Antonio Clericuzio, John Henry, Christoph Lüthy, William R. Newman and Catherine Wilson, has broadened the remit of atomic thought in early modern philosophy, revealing its presence in Aristotelian alchemy, Neoplatonism and Pythagoreanism in addition to a permeating Epicureanism. See, for example, Clericuzio (2000); John Henry, 'Atomism and Eschatology: Catholicism and Natural Philosophy in the Interregnum', *The British Journal for the History of Science*, 15.3 (1982), 211–39; Christoph Lüthy, 'David Gorlaeus' Atomism, or: The Marriage of Protestant Metaphysics with Italian Natural Philosophy', in *Late Medieval and Early Modern Corpuscular Matter Theories*, pp. 245–90; Christoph Lüthy, 'The Invention of Atomist Iconography', in *The Power of Images in Early Modern Science*, ed. Wolfgang Lefèvre, Jürgen Renn and Urs Schoepflin (Dordrecht, 2003), pp. 117–38; Newman, *Atoms and Alchemy*; and Wilson, *Epicureanism*.

[69] I quote from the expanded 1696 translation, pp. 187–88. The fifteenth-century Latin printing is *Laertii Diogenis Vitae et sententiae eorum qui in philosophia probati fuerunt*, trans. Ambrogio Traversari (Rome, 1472).

connection was strengthened further by the belief that atomism had an ancient Judaic origin that long pre-dated Leucippus and Democritus. A passing remark from Strabo led to the labelling of one Mochus the Phoenician, whose works were thought to predate the Trojan Wars, as the founding father of atomic theory.[70] Perhaps largely due to their similarity in name, various early modern philosophers linked Mochus to Moses, as one who was influenced by the work of the prophet and responsible for the transmission of (Moses') atomic philosophy to Pythagoras and the Greeks. This was the opinion of, amongst others, Theophilus Gale, Ralph Cudworth and Henry More.[71] For More and the earlier atomist Daniel Sennert, Mochus's rather tenuous relation to Moses went some way to sanctioning their work on corpuscular philosophy.[72]

Importantly for this book, the Mosaic claim to atomism was also a branch by which atomic ideas made contact with hermeticism and Neoplatonism, as the three ancient figures – the great Judaic prophet, the legendary Hermes Trismegistus (supposed by many to be Moses or one of his contemporaries) and the Phoenician proto-philosopher – blended into one. By the seventeenth century numerous philosophers had explored the possible place of atoms in a Neoplatonic cosmos. In contrast to the recent flurry of works on the Aristotelian contribution to early modern material philosophies, less attention has been paid to the significant role of Platonism

---

[70] Theophilus Gale claims that he learned of 'Moschus a Sidonian' and 'that he was the Author of the opinions of Atomes' from Strabo in *The Court of the Gentiles* (London, 1670), p. 59.

[71] See Gale, who claims that Mochus's 'Philosophie was nothing else, but the Historie of the Creation' and that he 'traduced his Natural Historie from the Historie of the Creation, written by Moses' (pp. 59–60). See also Ralph Cudworth, *The true intellectual system of the universe: the first part; wherein all the reason and philosophy of atheism is confuted; and its impossibility demonstrated* (London, 1678), p. 12; Henry More, *Conjectura Cabbalistica* (London, 1653), p. 82.

[72] Sennert writes: 'And this Opinion was a most ancient Opinion, and is now attributed to one Mochus a Phaenician, who is reputed to have flourished before the destruction of Troy'. Daniel Sennert, *Thirteen books of natural philosophy* (London, 1660), p. 446.

in shaping corpuscular theories.[73] The Neoplatonic connection was vital for the acceptance, where it occurred, of atomic physics and metaphysics in Christian theology. As Sarah Hutton has argued, while Platonism was not the driving philosophical method of the universities, it does not follow that it failed to shape early modern natural philosophy and metaphysics – she cites Bacon, Descartes and Leibniz as prominent examples of seventeenth-century 'moderns' who were influenced by Neoplatonic thought.[74] Leibniz's monads follow the particulate Neoplatonic theories of Nicholas of Cusa, Giordano Bruno and indeed Henry More, all of whom embraced spirituality in the atomic concept from its relation to the indivisible divine intellect.[75] More was especially attracted to the theory by observing sympathies between Neoplatonic understandings of the indivisible soul and the incorruptibility of the atom, similitudes that likewise appealed to his reader and contemporary Thomas Traherne. For the latter, the link with hermeticism was also important, as the hermetic emphasis on spiritual transcendence encouraged the devout enquirer to reach into the very origins and ends of things, the indivisibles that weaved the fabric of divine creation.[76] A central motive of this book is to tease out the vital relationship between Neoplatonic atoms – for there were many varying atomic theories associated with Neoplatonism – and the poem,

[73] For work that has focused on the relations between Platonism and atomic philosophy, see Jan Opsomer, 'In Defence of Geometric Atomism', in *Neoplatonism and the Philosophy of Nature*, ed. James Wilberding and Christoph Horn (Oxford, 2012), pp. 147–73. See also *Platonism at the Origins of Modernity: Studies of Platonism and Early Modern Philosophy*, ed. Douglas Hedley and Sarah Hutton (Dordrecht, 2008).
[74] Sarah Hutton, 'Introduction', in *Platonism and the Origins of Modernity*, ed. Hedley and Hutton, pp. 1–8 (p. 5).
[75] On relations between the thought of Cusanus and Leibniz, see Jan Makovský, 'Cusanus and Leibniz: Symbolic Explorations of Infinity as a Ladder', in *Nicholas of Cusa and the Making of the Early Modern World*, ed. Simon J. G. Burton, Joshua Hollman and Eric M. Parker (Leiden, 2019), pp. 450–84.
[76] I consider Traherne's hermeticism further in Chapter Two, pp. 110–13. See also Carol L. Marks, 'Thomas Traherne and Hermes Trismegistus', *Renaissance News*, 19.2 (1966), pp. 118–31.

a form of expression with which they were inextricably linked. Contrary to Marjorie Hope Nicholson's fears, these atoms rather repaired the Platonic circle of cosmic influence than break it.[77]

As the ancient, pagan atomic theories were reconciled with a Christian universe, they became more acceptable to seventeenth-century intellectual thought. Two prominent philosophers whose work on matter theories were crucial to the poets of this study – even though one of them cannot be dubbed an 'atomist' at all – were Pierre Gassendi and René Descartes. Gassendi, a French priest, was the most influential European figure in leading the attempt to combine Epicurean physics with Christianity.[78] He advocated the existence of atoms and void but made room for the Christian God in Epicurus's cosmos: according to Gassendi, the atomic system necessarily depended on a providential deity in order to function. His atoms were hard indivisibles endowed with activity, but not all particles were equal. Gassendi argued for the presence of 'semina rerum', special compound corpuscles.[79] If, as Antonio Clericuzio has commented, 'all molecules have some inherent degree of activity, semina are endowed with formative power and "programme", being responsible for the generation of minerals, plants and animals'. The difference is that these seeds, congregated from 'special' atoms, were 'formed by God at the beginning and ... endowed by God with a "programme"'.[80] With this theory Gassendi

---

[77] I reference the pioneering thesis of Marjorie Hope Nicolson in *The Breaking of the Circle*. Her introduction explains: 'The Circle of Perfection, from which man for so long deduced his ethics, his aesthetics, and his metaphysics, was broken during the seventeenth century. Correspondence between macrocosm, geocosm, and microcosm, long accepted as basic to faith, was no longer valid in a new mechanical world and mechanical universe' (p. xxi).

[78] For work on Gassendi, see Saul Fisher, *Pierre Gassendi's Philosophy and Science: Atomism for Empiricists* (Leiden, 2005); Lynn Sumida Joy, *Gassendi the Atomist: Advocate of History in an Age of Science* (Cambridge, 1987); Antonia LoLordo, *Pierre Gassendi and the Birth of Early Modern Philosophy* (Cambridge, 2006).

[79] Antonio Clericuzio introduces Gassendi's understanding of 'semina rerum' in *Elements, Principles and Corpuscles*, pp. 63–70.

[80] Clericuzio, pp. 68, 70 and 71.

avoided the trap of mindless, random materialism, as all corpuscular formations were preordained and managed, 'programmed' even, by the deity. As Margaret J. Ostler has explained, in Gassendi's universe 'the action of causes is simply the motion of atoms', but 'their mobility and activity function only with divine assent'.[81]

Gassendi wrote most of his philosophical works during the 1620s, 1630s and 1640s, but it was Walter Charleton, later a founding member of the Royal Society, who popularised his ideas in England with his patchwork *Physiologia Epicuro-Gassendo-Charletoniana: or a fabrick of science natural, upon the hypothesis of atoms* (1654).[82] Descartes also developed a corpuscular philosophy, but, unlike Gassendi, he rejected atoms and void. His reason dictated that truly indivisible particles of matter could not exist in a universe created by an infinite deity:

> For if there were any atoms, then no matter how small we imagined them to be, they would necessarily have to be extended; and hence we could in our thought divide each of them into two or more smaller parts, and hence recognize their divisibility. For anything we can divide in our thought must, for that very reason, be known to be divisible; so if we were to judge it to be indivisible, our judgement would conflict with our knowledge. Even if we imagine that God has chosen to bring it about that some particle of matter is incapable of being divided into smaller particles, it will still not be correct, strictly speaking, to call this particle indivisible. For, by making it indivisible by any of his creatures, God certainly could not thereby take away his own power of dividing it, since it is quite impossible for him to diminish his own power[.][83]

Within the Cartesian method all philosophy begins, in a sceptical fashion, with what can be proved by reason – if the reason can conceive of a minute but figured corpuscle being divided, then

---

[81] Margaret J. Ostler, 'How Mechanical was the Mechanical Philosophy? Non-Epicurean Aspects of Gassendi's Philosophy of Nature', in *Late Medieval and Early Modern Corpuscular Matter Theories*, pp. 423–40, p. 427.

[82] Walter Charleton, *Physiologia Epicuro-Gassendo-Charletoniana: or a fabrick of science natural, upon the hypothesis of atoms* (London, 1654).

[83] René Descartes, *Principles of Philosophy*, in *The Philosophical Writings of Descartes*, trans. John Cottingham, Robert Stoothoff and Dugald Murdoch, 2 vols (Cambridge, 1985), I, pp. 177–291 (pp. 231–32).

truly indivisible particles are an impossibility. Famously, in the *Discourse on Method*, the one principle Descartes proves as certain, according to his reason, is that there is an infinite, omniscient God to which all things owe their existence.[84] With this fundamental truth in mind, he argues in the *Principles of Philosophy* that it would be nonsensical to accept indivisible matter in a universe headed by an infinite power. This argument does not, however, reject the possibility that there is matter in the universe that will never be further divided. Descartes is willing to acknowledge that there may be material particles that remain, in actuality, indivisible, but they will never be truly indivisible as long as there is the potential – according to reason and the omnipotence of God – to separate them into smaller parts. He follows a similar process of reasoning in the care he takes to distinguish that which is *infinite*, and that we can only call *indefinite*; true infinity, Descartes argues, belongs to God alone.[85]

Of the poets featured in this book, Henry More, Margaret Cavendish and Thomas Traherne engage directly with Cartesian physics and metaphysics. Descartes's forays into the indefinite universe are an important influence on More's philosophical poetry, which goes one step further in accepting indivisibles as the essence of universal matter. The parameters of my study coincide with the two most influential philosophical *Principia* of the century – Descartes's *Principia Philosophiæ* of 1644 and Newton's *Philosophiæ Naturalis Principia Mathematica* of 1687.[86] These texts mark the philosophical background against which writers of the period conceived of the 'atom'. My endeavour in this book, however, is not to conduct a continuist study of the history of 'atoms' in seventeenth-century literature nor, primarily, to trace atomic theories in poetry back to

---

[84] Descartes, *Discourse on Method*, in *The Philosophical Writings of Descartes*, I, pp. 111–51 (p. 129).

[85] 'Our reason for using the term "indefinite" rather than "infinite" in these cases is, in the first place, so as to reserve the term "infinite" for God alone.' *Principles of Philosophy*, p. 202.

[86] René Descartes, *Renati Des-Cartes Principia philosophiae* (Amsterdam, 1644); Isaac Newton, *Philosophiæ naturalis principia mathematica* (London, 1687).

philosophical sources (though I shall be acknowledging the impact of these philosophies upon poetic works throughout, by necessity). While Robert Kargon, in his influential history of atomism in early modern thought, intended his work to address 'real men' rather than unspecific, abstract ideas, his 'real men' remained the elite scholars of the aristocracy and prestigious institutions.[87] Atomic thought spread much further than this. In addition to Kargon's philosophers, over these pages the reader will find clergymen and poets; not to mention a substantial number of 'real women' who contemplated atoms in philosophical 'fancies', biblical epics and devotional verses.

My first chapter considers the philosophical poetry of the Cambridge Platonist Henry More. In 1642, the young philosopher published his first writings, *Psychodia Platonica, or, A Platonicall Song of the Soul, consisting of Foure Severall Poems*. These early works were followed closely by another poetical composition, *Democritus Platonissans, or, An Essay upon the Infinity of Worlds out of Platonick Principles* (1646) and an expanded collection, *Philosophicall poems. A Platonick Song of the Soul; Treating, of the Life of the Soul, her Immortalitie, the Sleep of the Soul, the Unitie of Souls, and Memorie after Death* (1647). It was not until the mid-1650s that More started to publish the prose philosophical works with which he is most commonly associated today. More, unlike many of the writers of this study, was an atomist – but his atomism owed more to Pythagorean and Platonic metaphysics than to Epicurean physics and had its foundations in poetry. A fervent admirer of *The Faerie Queene* since childhood, More adopts the Spenserian stanza form as the location for his initial study of 'indivisibility' (a word he coined in 1647). The driving argument of the *Philosophicall Poems* is the need to know our inner selves in order to know divinity. More refers to the inner lives of things as 'Atom-lives', immanent and indivisible points of contraction, indebted to his understanding of the founding particle of physical and metaphysical being. My focus in this chapter is on these 'Atom-lives' and the easy, productive

---

[87] Robert Kargon, *Atomism in England from Hariot to Newton* (Oxford, 1966), p. vii.

intimacy between Platonic poetics and atomic thought in More's early writings. In the *Philosophicall Poems* More recognised that both divinely inspired forms, the poem and the particle at the base of creation, provided a means by which introspection of the immortality of one's own soul was made possible.

My second chapter turns to the Anglican priest and poet Thomas Traherne. For centuries, many manuscript texts of Traherne's oeuvre remained unknown to modern readers: it was not until Bertram Dobell's unearthing of a poetry manuscript in the 1900s and a series of discoveries at the close of the twentieth century (at the respective sites of the Folger Shakespeare Library, Lambeth Palace, and a Lancashire burning rubbish tip) that the greater extent of his scholarship was revealed. Traherne was very much influenced by Henry More – whom he cites and quotes from in his notebooks – and he derived his own atomic theory as a means for the obtainment of grace and 'Felicitie', his spiritually loaded term for divine bliss. Traherne's knowledge of and contribution to natural philosophy has long been understated by literary criticism. Following More and his own concoction of Aristotelianism and Platonism, Traherne embraced the atom as a receptacle of divinity. His theories moved beyond More's, however, by drawing a startling equivalence between the atom and the rational soul. With a reading of his reflections in the encyclopaedic (and recently discovered) works *The Kingdom of God* and *Commentaries of Heaven* (c.1673), I show how Traherne's atoms became models of behaviour for the soul on its quest to know and absorb 'All Things'. Poetry was the medium he adopted to facilitate this journey. His active, nimble poetic forms, themselves meditations on the divine 'Act' within all things, in turn accommodated atomic movements and grasped the manifold, wondrous possibilities of indivisibles at the heart of creation.

In the third chapter, the focus of the book shifts to the experimental poetry of two seemingly different writers: the obscure Hester Pulter and the more prominent seventeenth-century personality Margaret Cavendish. Both were inspired to write poetry on atoms, but while Cavendish's 'atom poems' from *Poems and Fancies* (1653) were intended for print and underwent three editions during her lifetime, Pulter's devotional poetry was composed in rural isolation and remained in manuscript. Pulter and Cavendish identified a

Introduction

creative power about the atom that fuelled their poetical, philosophical and theological ambitions. In this chapter, I explore the nuances of their dealings with the indivisible: their evocation of the atom for problem-solving and soothing; for extending the possibilities of poetic space; and for advancing their shared desire for a new world. For Cavendish, atoms play a significant part in her emphasis on the creation of worlds – familiar and new – in her imaginative writing; for Pulter, on the contrary, atoms and alchemy fuel the theological trust she places not in construction, but in dissolution. Atoms 'enfranchise' (to borrow one of Pulter's favourite words) the poetry of both women, but while Cavendish's early 'atom poems' enable her to 'build a new world in the mind' (*The Blazing World*, 1666), Pulter trusts in the dissolved 'indivisibles' of dust and decay to edge ever closer to the world's divine source.

Pulter's devotional trust in atoms paves the way for the subject of the fourth chapter, which explores the role played by the atom in Lucy Hutchinson's creation poem *Order and Disorder* (1679). The poem responded to *Paradise Lost* and was influenced by Hutchinson's full translation of *De Rerum Natura*, composed some twenty years earlier. While atoms in the universe of Milton's epic occupy undefined space at the margins of creation, the atoms of *Order and Disorder* provided Hutchinson with a means for understanding the ebb and flow of divine grace behind material structures. Nevertheless, while it has been acknowledged, criticism has been curiously reluctant to discuss the undeniable influence of Lucretius upon her creation poem. To claim that Hutchinson's theological poetics was inspired by atomic analogies and the movements of 'indivisibles' is not to claim that she was an atomist, or that she aligned her beliefs with Epicureanism. Instead, it is to tease out in her writing a deepening of empathy with mortal, physical existence, with the experience of oneself and others as creatures raised from matter – the understanding and communication of which are enabled by her atomic perspective of the fallen human condition. In this chapter, I explore this perspective from two, albeit intersected, directions. The first is from the portrayal of Hutchinson's Adam who, based on a background cultural understanding of Adam as the first principle, or 'atom' of mankind, is the first in *Order and Disorder* to understand the forthcoming pattern of corruption and regeneration. Secondly, and closely related to this depiction of the first man, I consider the significance of empathy in the poem's writing of postlapsarian bodies. This empathy, an appreciation of and effort to reconnect with first principles, leads to the promise of future resurrection, a purified renewal of being that Hutchinson expresses in atomic and chemical terms.

The subject of this book is a new understanding of the significant relationship between atoms and poetry in seventeenth-century England, a relationship that encouraged spiritual interpretations of the indivisible particle and revealed its intimacies with the origins of creation, human identity and God. My focus on this alternative poetics of the atom is part of a bigger mission to reconceive the complex relations between, not just poetry, but specific uses of poetic form and intellectual thought in early modern culture.[88] In a way parallel to Aït-Touati's recent choice of cosmological texts as 'the perfect material' for the study of 'this shared imagination' between the poetic and the natural philosophic, I focus my analysis on the 'perfect material' of the atom as it enters poetry.[89] Minute and unfathomable in and of itself, the concept of the atom takes on immense possibilities within poetic spaces that recognised the value of its indivisibility. The enormous potential of the atom – the indivisible at the core of matter; the building block of all things – captured the seventeenth-century poetic imagination and created new opportunities for it to 'pierce deep' (Bacon) into ontological subjects.

---

[88] Surprisingly little critical attention has been paid to the influence of poetic forms upon early modern literary engagements with natural philosophical themes, outside of the study of Lucretian tropes, though recent monographs by Elizabeth Scott-Baumann and Wendy Beth Hyman provide notable exceptions: Elizabeth Scott-Baumann, *Forms of Engagement: Women, Poetry and Culture 1640–1680* (Oxford, 2013); Wendy Beth Hyman, *Impossible Desire and the Limits of Knowledge in Renaissance Poetry* (Oxford, 2019).

[89] Aït-Touati, p. 7.

Chapter 1

ATOMIC CONGRUITY: THE PHILOSOPHICAL POETRY OF HENRY MORE

Since beginning work on More as a graduate student, I have been asked several times why I choose to write about 'bad poetry'. The question may be tongue-in-cheek, but it rests on a critical commonplace that has long disheartened readers of More's early philosophy. More's prose has a level of elegance that is lacking in his 'conspissate' Spenserian stanzas, where the inclusion of words like 'swonk' did not exactly help his cause.[1] As early as 1650, only eight years after the first publication of the Cambridge Platonist's verse, the alchemist Thomas Vaughan (no fan of More's) dismissed the author as 'a poet in the loll and trot of Spenser', a mindless imitator of outdated poesy.[2] In 2016, Guido Giglioni reviewed a new study of More's metaphysics by Jasper Reid and made the following insight:

[1] More uses the Anglo-Saxon 'swonk' (the past tense of 'swink', meaning to toil or to work hard) in his first collection of poems, *Psychodia Platonica: or, a Platonicall Song of the Soul* (Cambridge, 1642), sig. F3. 'Conspissate', meaning to thicken or to make dense, is a neologism by More that appears first in his *Democritus Platonissans, or, An essay upon the infinity of worlds out of Platonick principles* (Cambridge, 1646), sig. B2v.
[2] 'To my Learned, and much Respected friend, Mr. Mathew Harbert', in *The Man-Mouse Taken in a Trap, and tortur'd to death for gnawing the Margins of Eugenius Philalethes* (London, 1650). Quoted from *The Works of Thomas Vaughan*, ed. Alan Rudrum (Oxford, 1984), p. 237.

> More's philosophical inquiry feeds on experiments, stories, and visions, and these become an organic part of his metaphysical equipment. This, of course, was a hazardous and edgy undertaking, which no doubt contributed to the bad reputation associated with the style of More's philosophizing. Perhaps due to his poetic vein (the question of its aesthetic value is not relevant here), More was prone to linguistic exuberance.[3]

Giglioni's observation is an illuminating one for those embarrassed by More's poetry. He claims that More's philosophy 'feeds on' creative hypothesis and fiction: there was something about his early poetic impulse that was essential to the growth of his metaphysics. Giglioni's follow-up point, however, that this 'poetic vein' resulted in 'hazardous' effects of 'linguistic exuberance', omits half of the story. More's 'linguistic exuberance' was not a side-effect of an over-active imagination, but a purposeful style he considered essential for philosophical communication. It was also a key drive behind his need for poetry. The coining of new words was an inseparable part of the poetic act, and wordiness a requisite of actualising theories about lofty matters. It is telling that, in Giglioni's review, the supposedly dubious quality of More's poetry is once again acknowledged; but why must the question of aesthetic value be irrelevant?

More's early publications forge a link between poetic form and the founding of philosophical theories. Specifically, and crucially for this study, there is an important connection between More's experimental poetics and the construction of a hybrid Platonic–Epicurean atomism. His poetic influences – Spenser, devotional poetry, hermetic and Orphic song, Lucretius – combined to develop an 'aesthetic' that prepared not only the verbal representation, but the realisation of otherwise unfathomable entities in his universe. Letting the fancy roam was a pivotal stage of the philosophical process for More, who argued that all advances in knowing were reliant on self-reflection and expression. For More, the poet was focused on the nature of mind and self, on his in(-)dividuality in two senses of the word: on the one hand, his separation from the Neoplatonic 'World Soul' and recognition of this individual

---

[3] Guido Giglioni, 'The Metaphysics of Henry More by Jasper Reid (review)', *Journal of the History of Philosophy*, 54.3 (2016), 502–03 (502).

fragmentation; on the other, his awareness of the universal origin and his aspiration to return there (to return to in-dividualism and put an end to division). Self-reflection became a means to unearth the buried glory of the fallen, embodied rational soul. The power of poetry regarding this worked both ways: poetic expression facilitated engagement with the inner life of the self, but its method was to appeal to and absorb divine inspiration. More's poetry therefore rests on the paradox that the more we reflect on the nature of being human, the more we discover the details of the divine universe beyond human comprehension.

The atom is the key to unlocking this paradox, because it encapsulates it. As an unfathomable component of being, the atom assumes qualities when it becomes the subject of poetry, as it does so frequently in More's *Philosophicall Poems*. More than that, however, the indivisible particle, like poetry and like human beings, stems in part from the material world – the experience of embodiment – and in part from its spiritual origins and ends. It is by knowing our inner selves, More argues, that we shall know divinity – and he refers to the inner lives of things as 'Atom-lives'; immanent and indivisible points of contraction.[4]

The first section of this chapter introduces some of More's key philosophical theories surrounding matter and spirit, notably his dedication to the presence of 'indivisibility', a word he coined in his early poetry.[5] His language of indivisibility blurs the physical and the spiritual, an ambiguity that corresponds with his quasi-materialist understanding of, to quote Plotinus, 'The Soul's Descent into Body'.[6] More's theory of 'Vital Congruity' – the fit between soul and matter – feeds into his understanding of the founding particle of

---

[4] More defined 'Atom-lives' in 'The Interpretation Generall' which accompanied the first and third collections of his poetry, *Psychodia Platonica* in 1642 and the expanded *Philosophicall poems* (Cambridge, 1647). See my analysis of the definition later in this chapter, p. 54.
[5] More coins the word 'indivisibility' in *Psychodia Platonica* (see p. 53).
[6] *Plotinus: The Enneads*, ed. John Dillon, trans. Stephen MacKenna (London, 1991), The Fourth Ennead, Eighth Tractate, p. 338. All further references to *The Enneads* will be taken from this edition and cited in the text by page number.

metaphysical being: the atom, which he later renames the monad. The change of terminology is not a rejection of his former theory, but a realisation of the atom's power as a model of contracted indivisibility – an essential source for understanding the origin of self and thence the immanence of God. Following this introduction to More's ensouled atomism, the second section turns more directly to the (literally) vital importance of poetry for the exploration of 'Atom-lives'. In the *Philosophicall Poems* the atom becomes not merely a singular of the particles that constitute the body, but an emblem for the vitalist impulse of knowing self and the divine.

MORE'S PHILOSOPHY AND ATOMS

In the *Philosophicall Poems*, More sought to merge what would appear to be philosophical polar opposites: Neoplatonic emanationist theory and Epicurean materialism.[7] His combination of the schools was motivated in part by a desire to rewrite philosophical history. He considered Pythagoras, before 'Epicurus, Democritus, Lucretius', to be the father of atomism and subsequently a figure who formed a bridge between Neoplatonic spiritualism and materialism.[8] This methodology is implied in More's epistle 'To the Reader' at the front of the *Philosophicall Poems*, where he explains that he has 'added Notes for the better understanding … of the Principles of *Plato's* Philosophy' and 'beside some subtil considerations concerning *ATOMS* and *QUANTITY*, set out very plainly, the *Hypothesis* of *Pythagoras*, or *Copernicus* concerning the *MOTION* of the *EARTH*'.[9] More's early writings present an easy pairing of Platonic poetics and atomic thought. Emanationist cosmogonies, as John Henry has summarised, 'always bear a materialistic

---

[7] For more on the incompatibilities between the Platonic world and Democritean atomism, see Catherine Wilson, *Epicureanism at the Origins of Modernity* (Oxford, 2008), pp. 45–47.
[8] More makes this point in his preface 'To the Reader' in *Democritus Platonissans*, sig. A2.
[9] See *Philosophicall poems*, sigs B2r-v. See also A. Rupert Hall, *Henry More: Magic, Religion and Experiment* (Oxford, 1990), p. 111 (on More's understanding of the Pythagoreans as 'the founders of Greek atomism').

stamp because the natural world is held to consist of a series of increasingly crass and crude emanations from the Godhead'.[10] The dense matter at the bottom of the scale is the most distant from the original divine light. Moreover, though More insisted on the immateriality of spirit, he endowed it with characteristics typical of material entities, including a 'Plastick' power that grouped substances together and fitted bodies to souls.[11] He was so eager to establish the pervading 'spirit' of the world as a medium between mundane and celestial forms of experience that he assigned it extension and density. Despite a proposed dualism in his thought, therefore, there is often little to distinguish between the characteristics of material and immaterial substances in his writing.[12]

The details of More's 'dualism', though he repeatedly asserts the distinction between matter and spirit through their opposite qualities, remain consistently unclear. Ultimately, the Neoplatonist could not subscribe to a mechanical universe where material actions were mindless and soul-less, though he was initially taken by the work of the most famous contemporary dualist of all: Descartes. In 1647, five years following the initial publication of his *Psychodia Platonica* and three years after the emergence of Descartes's *Principia*, More returned to his collection of philosophical poetry to make significant amendments. In terms of methodology, More was struck by Descartes's internalised God and mind-driven approach to philosophy, so concordant with his own appeals to the immanence of the divine in human psychology. He similarly found much to occupy his thought in the Cartesian theories of space and extension and in the possibility of an indefinite universe. More appreciated Descartes's understanding of the passivity of matter and the

---

[10] John Henry, 'A Cambridge Platonist's Materialism: Henry More and the Concept of Soul', *Journal of the Warburg and Courtauld Institutes*, 49 (1986), 172–95 (179).

[11] More first refers to the 'Plastick might' of the 'spirit of life' in *Psychodia Platonica*, specifically in his poem *Psychathansia*, Book 2, Canto 1, stanza 9 (p. 36).

[12] For more on this, see Stephen Clucas, 'Poetic Atomism in Seventeenth-Century England: Henry More, Thomas Traherne and the "Scientific Imagination"', *Renaissance Studies*, 5.3 (1991), 327–40 (337).

active nature of spirit, but his later, more sceptical works rejected mechanical philosophy.[13] He wrote to Descartes in 1649 to expose the flaws in Cartesianism and to contend that 'there are phenomena that cannot be explained mechanically'.[14] One notable point of difference was in how the philosophers respectively defined the corpuscular components of the material world. While Descartes rejected the reality of atoms, with the claim that anything divisible according to the imagination cannot, by consequence, assume true indivisible status, More argued for the necessity of indivisible constituents for which further division would be not just a physical, but a metaphysical impossibility. His *Philosophicall Poems* are firm in establishing this. In 1647, More coined the word 'indivisibility': he was, remarkably, one of the first (if not *the* first) English writers to consider the 'indivisible' not merely as an adjective but as a general state of being.[15]

The metaphysical absolute of indivisibility is a cornerstone of More's philosophy. In the *Philosophicall Poems*, indivisibles are irreducible. Even God – the creator of all things – cannot reduce that which is indivisible, though he reserves the power to uncreate. A decade later, however, the wording of More's compressed material philosophy and metaphysics changed. More made the shift to his enduring preference for 'indiscerpibility', a word coined

[13] For more information on More's changing attitudes to Cartesianism and metaphysics, see Alan Gabbey, 'Henry More and the Limits of Mechanism', in *Henry More (1614–1687): Tercentenary Studies*, ed. Sarah Hutton (Dordrecht, 1990), pp. 19–35.

[14] More's letter to Descartes from 23 July, 1649. Quoted from Stuart Brown, 'Leibniz and More's Cabbalistic Circle', in *Henry More: Tercentenary Studies*, pp. 77–96 (p. 80).

[15] The first recorded use of 'indivisible' as noun predates More's revised collection of the *Philosophicall Poems* only by three years, in the Aristotelian atomist Kenelm Digby's *Two treatises in the one of which the nature of bodies, in the other, the nature of mans soule is looked into in way of discovery of the immortality of reasonable soules* (London, 1644), p. 462. Interestingly, Digby's reference is to an 'indivisible' instant of time, in relation to the 'permanent Eternity' of 'this great word Forever'.

in *The Immortality of the Soul* (1659).[16] Ironically for a word so heavily compounded, that which is 'indiscerpible' is incapable of being divided further into parts. From 1659 onwards, More's replacement word was predominantly associated with the nature of soul and spirit. The word accommodates the at-one-ness of soul, as More explains in one of his most accessible works, *Divine Dialogues* (1668):

> By the *Self-unity* of a Spirit I understand a Spirit to be *immediately* and *essentially one*, and to want no other *Vinculum* to hold the parts together but its own essence and existence; whence it is of its own nature *indiscerpible*.[17]

An 'indiscerpible' substance, therefore, is independent and self-contained; it is 'immediately and essentially one', so that if any change were to come upon it, the whole entity would necessarily be affected uniformly. This is complicated by More's claim that spiritual entities (souls) have extension. According to *The Immortality of the Soul*, body is 'a substance impenetrable and discerpible' while its contrary, spirit, is 'a substance penetrable and indiscerpible' (p. 17). For these differences, he argues, it is important to remember that 'the precise notion of Substance is the same in both, in which, I conceive, is comprised *Extension* and *Activity* either connate or communicated' (p. 17). More has carried over the vocabulary of atomic matter (*Extension, Activity*) to his exploration of indivisible spirit, with the result that the notion of an 'indiscerpible', extended substance might more instinctively evoke thought of a physical particle, not a human soul.

---

[16] This is recorded in the *OED*, 'indiscerpibility, n'. The first usage is from *The immortality of the soul, so farre forth as it is demonstrable from the knowledge of nature and the light of reason* (London, 1659), where More writes of 'the *motion* of a Soul, which is a *Spirit*, and therefore of an *Indivisible*, that is of an *Indiscerpible*, Essence' (p. 241). All further quotations from this work will be cited by page number in the main text.

[17] *Divine dialogues, containing several disquisitions and instructions touching the attributes of God and his providence in the world* (London, 1668), p. 124. All further quotations from this work will be cited by signature or page number in the main text.

The attribute of extension understandably raises confusion amongst the characters of the *Divine Dialogues*. More's dialogue is led by the wisest and worthiest participant, Philotheus, introduced in the prefatory material as 'a zealous and sincere Lover of God and *Christ*, and of the whole Creation' (sig. b4v). Unsurprisingly, Philotheus is the character who most keenly represents the thoughts and conclusions of the author himself. On the topic of matter, spirit and extension he engages closely with the opinions of Cuphophron, 'a zealous, but Airie-minded, *Platonist* and *Cartesian*, or *Mechanist*', and Hylobares, 'a young, witty, and well-moralised materialist' who, initially bound only to material causes and consequences, turns to God over the course of the evening's conversation (sig. b4v).[18] Cuphophron is committed to the writings of Descartes, accusing Philotheus and Hylobares of plotting 'a conspiracy together against that Prince of Philosophers' (p. 122). If Philotheus is the mouthpiece of the mature More, Hylobares recalls his younger counterpart of several decades previously, who is/was 'well-moralised' and receptive to new learning and divine truth. Hylobares responds to his Socratic teacher/future self by offering a definition of spirit based on what he has derived of its opposite, matter. If '*Self-disunity, Self-inactivity, Self-impenetrability,* be the essential Attributes of *Matter* or *Body*; then the Attributes of the opposite *species,* viz. of *Spirit,* must be *Self-unity, Self-activity, Self-penetrability*' (p. 124). He struggles, however, with the consequence of self-unity – indiscerpibility. For how, Hylobares asks, 'can an extended Substance be indivisible or indiscerpible?' (p. 124) The mature More's character replies with the following:

> It is true, it is *intellectually* divisible, but *Physically* indiscerpible. Therefore this is the fallacy your Phancy puts upon you, that you make *Indivisibility* and *Indiscerpibility* all one. What is *intellectually* divisible may be *Physically* indivisible or indiscerpible: as it is

---

[18] *Divine Dialogues* concludes with Hylobares's closing joy in the 'Harmonie of God': 'I am not onely at perfect ease touching all Doubts about Divine Providence, but in an ineffable Joy and Ecstasie, rapt into Paradise upon Earth, hear the Musick of Heaven, while I consider the Harmonie of *God,* of *Reason;* and the *Vniverse,* so well accorded by the skilfull voice of *Philotheus.*' *Divine Dialogues*, p. 550.

manifest in the nature of God, whose very *Idea* implies Indiscerpibility, the contrary being so plain an Imperfection. (p. 125)

More refers to spiritual substance, but the terminology he uses to denote its features sets the distinction between matter and spirit in doubt. The qualification 'physical' is applied to the indiscerpibility of extended spirit. Philotheus's purpose is to explain that spirit is inaccessible to material acts of separation, not to claim that it has physical properties – but he muddies the waters by relying on physical vocabulary to get there. He refutes the common Cartesian argument against indivisibility by distinguishing between division and his new term, discerpibility. They are not synonyms, because a substance might be divided by the imagination but that does not mean it can be separated actually into parts, which is equated here with a physical act. Spiritual substances, though extended, differ from material substances because they cannot be broken down into smaller bits. It is nevertheless striking that Philotheus's reasoning relies on bringing physical experience to spiritual entities. Spirit – even God – is extended and indiscerpible, but obviously capable of physical interaction and even defined in relation to physical characteristics: that which it is not. When Philotheus denigrates 'intellectual' conceptions of divisibility his argument comes into conflict with a canon of metaphysics that would deem 'intellect' the purest, most sophisticated capability of the rational soul. More asks us to return to the language of physics to find the answers to the biggest questions in metaphysics and divinity.

The suggestion here is that spirit is dependent on physical properties not just for the distinguishing of its substance – its existence as that which is *not* body, *not* divisible, *not* dependent – but for the very enabling of its being according to mortal experience. This links with one of the boldest features of More's metaphysics, the theory of 'Vital Congruity'. In a controversial move that subjected him to criticism from scholarly contemporaries, More argued that souls necessarily require the material form of a vehicle in order to become active.[19] The union he describes between body and soul is

---

[19] More's most notable critics were Richard Baxter and Robert Boyle. See Henry, 'A Cambridge Platonist's Materialism', pp. 183–88, and 'Henry More

not merely mechanical, but 'vital'. If it were mechanical, he explains in *The Immortality of the Soul*, the soul would be imprisoned inside a corporeal cell; as it is, however, the soul has the ability to move and penetrate other bodies according to her 'Will' and 'Imagination' (p. 275). As the soul progresses in its journey between the terrestrial and celestial worlds, the matter of its vehicle transforms to become more subtle and refined. Immediately following the demise of the mortal body, it ascends into the air and adopts the clothing of an 'aerial vehicle', in which state it may remain for 'many Ages'. The third and final stage of transcendence occurs when the soul rises into an 'aetherial' or 'coelestial' vehicle, which it inhabits 'forever' (p. 158). The further the soul travels into more subtle degrees of matter, the more divine its capabilities to act. More provides an explanation in *The Immortality of the Soul* to defend his reasons for attaching vehicles to souls:

> 8. ...It is plain therefore, that this Union of the Soul with Matter does not arise from any such gross Mechanical way, as when two Bodies stick one in another by reason of any toughness and viscosity, or straight commissure of parts; but from a *congruity* of another nature, which I know not better how to term then *Vital*: which *Vital Congruity* is chiefly in the Soul it self, it being the noblest Principle of Life; but is also in the Matter, and is there nothing but such modification thereof as fits the *Plastick* part of the Soul, and tempts out that Faculty into act.
>
> 9. Not that there is any *Life* in the *Matter* with which this in the Soul should sympathize and unite; but it is termed *Vital* because it makes the *Matter* a congruous Subject for the Soul to reside in, and exercise the functions of life. (p. 263)

More's expression 'Vital Congruity' emerges in response to a state of being that he knows 'not better how to term'. Though he insists throughout his writings on a strict dichotomy between corporeal matter and soul, the division is undermined when he considers how matter and spirit unite and asks why. The 'Plastick part of the Soul' acts as go-between by preparing the matter

---

versus Robert Boyle: The Spirit of Nature and the Nature of Providence', in *Henry More: Tercentenary Studies*, pp. 55–76.

as a place of inhabitation; the metamorphic vehicular extension of the soul then shifts in accordance with its surrounding environment. What the 'Plastick' part consists of and the location it inhabits within the soul remain unclear.[20] It creates the 'fit' between matter and soul, More proceeds to explain, by means of an 'unresistable and unperceptible pleasure' (p. 265). His negative words respond to and, in doing so, increase a certain ambiguity – is the pleasure experienced by the soul irresistible because its spiritualisation of matter brings so much joy, or has the soul been forced downwards into a physical union it is unable to refuse? Similarly, is the pleasure 'unperceptible' to terrestrial mankind or to the soul itself?

More's understanding of the descent of the soul is largely inspired by Plotinus's discussion of 'The Soul's Descent into Body' in the fourth Ennead (pp. 334–43). Plotinus explains that individual human souls emerge from the universal 'All-Soul' when they become 'partial and self-centred', with 'a weary desire of standing apart' (p. 338). The fall into the corporeal world is a fall into temptation, and as a result the soul loses its intellectual powers: 'the Soul is a deserter from the totality; its differentiation has severed it; its vision is no longer set in the Intellectual; it is a partial thing, isolated, weakened, full of care, intent upon the fragment' (p. 338). In *The Immortality of the Soul*, More's imagery suggests that he similarly views the soul's journey into congruity with matter as a descent into an inferior state of being, triggered by inferior and partial desires. Tempted into incarceration, the descent of the soul is likened to a hubristic fall. Like 'Eagles' and 'Birds of prey', the soul 'may be fatally carried, all Perceptions ceasing in her, to that Matter that is so fit a receptacle for her to exercise her efformative power upon' (pp. 265–66). The predator-and-prey analogy harks back to a similar image from the early poem *Psychathanasia*. More develops his theories for the spiritual formation of the 'lower man', explaining that 'hovering souls' are

---

[20] For more information, see Henry, 'A Cambridge Platonist's Materialism'. D. P. Walker discusses some of the materialist characteristics of More's 'Spirit' in 'Medical Spirits in Philosophy and Theology from Ficino to Newton', in *Arts du Spectacle et Histoire des idées* (Tours, 1984), pp. 287–300.

attracted to matter 'When it's right fitted'. As a consequence of this desire, 'those spirits fall / Like Eagle down to her prey, and so endure / While that low life is in good temperature'.[21]

If the above-quoted chapter from *Immortality* is taken in its full context, however, the interpretation that the soul loses intellectual strength from its descent becomes restrictive and invalid. More stresses the lack of will in the soul's fall, but also the utter necessity that it meet the matter to which it is involuntarily drawn in order to exercise its 'efformative power'. He describes the body that is formed by the soul as a 'receptacle', a vessel for spirit that exceeds the limitations of a physical container. In addition to carrying the meaning of something that receives and gives admittance, a 'receptacle' bears the further connotations of offering sanctuary or providing shelter.[22] From this perspective, the soul is nurtured by the matter that receives it following its descent. Its amalgamation into the material world both enables the beginning of its ascent back into celestial unity and facilitates the growth and maintenance of the cosmos. The contrast between activity and passivity in More's imagery convincingly recalls Plotinus's assertion, also in 'The Descent of the Soul', that its communion with matter is 'a voluntary descent which is also involuntary' (*Enneads*, p. 339). If the soul itself cannot reason its descent into materiality, its movements must form part of an overall divine plan.

What, then, is the link between More's theory of vital congruity and his atomism? At first glance there would appear to be tensions between the two theories involving matter. Following his introduction to vital congruity in *The Immortality of the Soul*, More hastily insults what he sets out to be the contrary hypothesis of the atomists: 'this is one Hypothesis [vital congruity], and most intelligible to those that are pleased so much with the opinion of those

---

[21] *Psychathanasia*, Book 2, Canto 1, verse 10, p. 108 (quoted from *Philosophicall poems*). Further quotations from More's poetry will be taken from the composite collection *Philosophicall poems* (1647), unless stated otherwise. The poems *Psychozoia* and *Psychathanasia* will be cited by book (*Psychathanasia* only), canto and stanza number; other poems will be cited by page number.

[22] *OED*, 'receptacle, n'.

large Sphears they conceive of *emissary Atomes*' (p. 266). 'Emissary' implies that, according to this theory of materialist creation, atoms are up to no good. The information they seem to provide is dubious, even odious, as to conceive of physical objects and beings created by the mechanical motions of atoms alone smacks of atheism, recalling the suspicion of mindless, 'senseless' particles prevalent in contemporary literature.[23] Soul must be involved in the creation process and in the movements of bodies thereafter. In *Divine Dialogues*, Philotheus advises Hylobares to take leave of his materialism, explaining that matter requires direction from a 'higher Principle' – souls and World-Soul – to 'keep it in that consistence it is', because '*Disunity* is the natural property of Matter' (p. 32).

This rejection of materialism does not, however, banish indivisible particles from More's universe, nor diminish the significance of the atom in his metaphysics. More remained an atomist, but not of the Epicurean school. He was adamantly opposed to key features of Democritean–Epicurean atomism, among them the theory of vacuum (he argued instead for a plenum) and, as mentioned above, the notion that perceptive, mobile creatures could be formed by the random motion and collision of atoms.[24] For More, it was of *vital* importance – literally – that matter be ensouled so to be made active. While he maintained atoms in his cosmos, his verbal conceptualisation of the indivisible particle underwent change as he developed his theories of congruity. From 1662 he preferred the term 'physical monad'.[25] More has been acknowledged as the first English writer to use the word 'monad' when referring to the divine

---

[23] For just one example, see the classicist Richard Bentley, who dismisses 'this sottish opinion of the Atheists; That dead and senseless Atoms can ever justle and knock one another into Life and Understanding'. *The Folly and Unreasonableness of Atheism* (London, 1699), p. 56.

[24] More writes in his preface 'To the Reader', at the front of *Democritus Platonissans* (1646): 'But if any space be left out unstuffd with Atoms, it will hazard the dissipation of the whole frame of Nature into disjoynted dust' (sig. A2).

[25] More preferred this term from 1662 onwards, following its first usage in his 'Appendix to the Defence of the Philosophick Cabbala' from *A collection of several philosophical writings* (London, 1662). Jasper Reid makes this connection in *The Metaphysics of Henry More* (Dordrecht, 2012), p. 51.

unity of God (in *Psychodia Platonica*, 1642); his use of the word to denote an indivisible unit in philosophy, however, goes unrecorded by the OED, where the progress of the noun is traced from Giordano Bruno to Leibniz to Anne Conway, without mention of the Cambridge tutor.[26] In *Divine Dialogues*, Philotheus follows up his claim that *'Disunity* is the natural property of Matter' by explaining that it [matter] 'of it self is nothing else but an infinite Congeries of *Physicall Monads*' (p. 32). More's use of the word 'monad' here refers to atomism in everything but name. These monads are qualified as 'physicall' and as an infinite mass or heap of particles. They are indivisible units.

Crucially, physical monads are not the only indivisible units at work in the universe. More was exploring the significance of the 'monad' as early as 1642, when he referenced the term in 'The Interpretation Generall', the closing glossary, of his *Philosophicall Poems*:

> Monad. Μονάς, is Unitas, the principle of all numbers, an embleme of the Deity: and the Pythagoreans call it Θεός, God. It is from μένειν because it is μόνιμος, stable and immovable, a firme Cube of it self, One time one time one still remains still one (p. 430)

The monad, the first, unified principle, is defined as an 'embleme' of God. More appears at first glance to be distinguishing his 'monad' from its Greek source, as he is careful to identify what the Pythagoreans called 'God' with what he calls a symbol of the unmoved mover. The relation of his symbol to the origin of all things is nevertheless called into question by its status as a first principle. More bestows the divine monad with extension, just as he would soul and God in later works; it is a perfectly self-contained and independent 'firme Cube'. By cube, he refers directly to the Pythagorean tripling of this first principle of numbers, which remains constant under multiplication ('one time one time one still remains still one.') A Pythagorean 'Cube' can also, however, allude to the solid

---

[26] It is curious that an author so renowned for neologism and the Anglicisation of Greek vocabulary should be overlooked, especially when his first usage of the word in another context – and indeed of the adjective 'monadical' in his philosophical poetry – have been dutifully recorded. See *OED*, 'monad, n.' and 'monadical, adj.'

mathematical figure from which the element of earth derives its source, as was cited by the *Timeaus* ('to earth, then, let us give the cube') and pseudo-Plutarch in *Placita Philosophorum*: 'the universe is made from five solid figures, which are called also mathematical; of these he [Pythagoras] says that earth has arisen from the cube, fire from the pyramid, air from the octahedron, and water from the icosahedron'.[27] More's definition is worded so that, despite the primary intention to prove that the cubic root of the monad is 'of it self', his language leaves the impression of a solid, physical object. There is little to prevent a transferral of the above understanding of the divine to the atomic 'Physicall Monad' outlined in *Divine Dialogues*. His use of the word 'embleme' contributes to this part-materialist curiosity. An emblem is a symbol that represents an abstract concept, but it can also be an entity that in and of itself demands deeper interpretation. Some objects are by their very nature designed to be figurative, to embody multiple possible meanings, functions or possibilities. In other words, this self-contained indivisible – 'the principle of all numbers' – could well take form as an allegory of itself.

More's metaphysics are wrapped in dense, material paradoxes such as these, which find their natural home in the thought experiments of his *Philosophicall Poems*. The shift in his vocabulary from the atom to the monad was a move to a word he considered better suited to accommodating the varying substance, from physical to spiritual, of indivisibles in his cosmos. There is a significant and purposeful ambiguity between spiritual and physical things that is heightened in the presence of the atom or monad, stemming from an atomic perspective on soul and vitality that was developed first in More's early poetry. His strange, uncertain symbolism of the monad can be enlightened, at least partially, by an understanding of his vitalism. Christ's College Cambridge became a hub for vitalists in the seventeenth century. More matriculated at the college

---

[27] *Plato: Complete Works*, ed. John M. Cooper (Indianapolis, 1997), *Timaeus*, trans. Donald J. Zeyl, pp. 1224–291 (55e, p. 1258). Pseudo-Plutarch, *Placita Philosophorum*, Book 2, Chapter 6. Quoted from *Plutarch's Morals. Translated from the Greek by several hands*, ed. William W. Goodwin (Cambridge, MA, 1874), p. 135.

while Milton was still in residence and he worked alongside the experimentalist Henry Power, an atomist and follower of William Harvey, within the fellowship.[28] While More was keen to distinguish the 'pure potentiality' of unformed, inanimate matter from the superior spirit that enlivened it, he subscribed to the theory that life is produced in material beings by a vital principle. His theory of 'Vital Congruity' alone demonstrates that matter requires spirit to become active, but the relationship works both ways. In *Psychathanasia* he describes the 'Unmoved Monad' of the divine:

> One steddy Good, centre of essencies,
> Unmoved Monad, that Apollo hight,
> The Intellectuall sunne whose energies
> Are all things that appear in vitall light,
> Whose brightnesse passeth every creatures sight,
> Yet round about him stird with gentle fire
> All things do dance (3.3.12)

More's subject here is the 'Archtype' (3.3.7) of the light we see, the 'Intellectuall sunne' which 'we call God' (3.3.11) – in this context, the 'Unmoved Monad' is equated with a pervading, emanating Holy Spirit. While we may not directly see this 'Intellectual' source, we experience the consequences of its influence – its 'energies' – which are nothing short of 'all things that appear in vitall light'. The imagery here is classically Neoplatonic, with a vitalist perspective on the energising of material things. More's note on 'Energie' in

---

[28] More matriculated in 1631; Milton graduated from Cambridge in July 1632. See Sarah Hutton, 'More, Henry (1614–1687)' in *Oxford Dictionary of National Biography* <https://doi.org/10.1093/ref:odnb/19181> [accessed 12 Nov. 2019] and Barbara K. Lewalski, *The Life of John Milton: A Critical Biography* (Oxford, 2000), p. 53. See also Sarah Hutton, 'Mede, Milton, and More: Christ's College Millenarians', in *Milton and the Ends of Time*, ed. Juliet Cummins (Cambridge, 2003), pp. 29–41. Henry Power, a vitalist and atomist, was also, like More, a keen neologist: he is the 671st most cited source for word origins in the *OED*, even though he only published one book: *Experimental philosophy, in three books containing new experiments microscopical, mercurial, magnetical: with some deductions, and probable hypotheses, raised from them, in avouchment and illustration of the now famous atomical hypothesis* (London, 1664).

'The Interpretation Generall' is one of the longest in the collection, opening with the disclaimer that it is 'a peculiar Platonicall term' and attempting a definition through explaining its co-dependent relationship with 'Essence'. He describes the latter as 'the Centre as it were, of that which is truly called Energie, and Energie the beams and rays of an essence'. The origin of all things, the 'Intellectual sunne' or 'Unmoved Monad', provides the source of essence and energy but is not, More's 'Interpretation Generall' tells us, the sole monad in the universe. We are encouraged to look to ourselves. This he takes from Plotinus: 'For every being hath its Energie, *which is the image of it self*, so that it existing that Energie doth also exist' (italics mine, p. 426). More's repetitive wording highlights the reflective status of this internal energy, of which external energy is a mirror. He dubs this self-image, the 'depth, or inmost Being of any thing', with the label 'Centrall' or 'Atom-life' – the 'indivisibility of the inmost essence it self'. The quasi-allegorical, material paradoxes of More's monads and atoms begin to take on life from this comment. Atoms and essences conflate in this vocabulary and the model of contracted indivisibility – whether in the form of an atom or a monad – becomes vital for an understanding of self, others and relations with the divine.

These 'peculiar Platonicall' terms, unfathomable in isolation and co-dependent for any metaphysical understanding, were initially explored by More in poetic form. In the following section, I consider the significance of More's poetic expression in conceiving atoms and 'Atom-lives': the importance of his use of Spenserian form as the location for spiritual self-reflection and, consequently, the realisation of indivisibility.

VITAL POETICS

The very word 'poetical' has a vitalist impulse. From 'poesis', it implies something seminal and formative; something in the making. More's emblem for this impulse, through which the poet-creator imitated the divine maker, was the 'atom' or 'monad' – the indivisible point(s) he understood to be at the centre of all things, 'centreities' of concentrated essence that housed individual and divine identity. This association is, to apply an appropriate word, a little dense to say the least. As I explored in the previous section, More's understanding of the cosmos is inherently emblematic. His is a Neoplatonic universe of correspondences and symbolism but, moreover, all physical things come into being as enlivened allegories of themselves. Material entities take form as receptacles of immanent spirit, emblematic but, crucially, vital representations of the spiritual essence within. The cosmic structure is supported by a constant correspondence between the inner and outer realities of things, what Plotinus referred to as the energy of every being, '*which is the image of it self*, so that it existing that Energie doth also exist' (italics mine). Continual self-reflection is the key to created things taking form as the most fitting, and true representation of the concentrated, indivisible centre at the essence of their being. Humans, as possessors of a rational soul, are superior in the acts of self-knowledge and vital congruity. They can ascend through the layers of emanation towards the essential light, their vehicle assuming purer substance, thus sharing greater intimacy with their spiritual centres on their rise. From the very beginning of his philosophical career, More associated the centres of enlivened things with atoms, naming them 'Atom-lives'.

'Atom-lives', otherwise known as 'Centrall-lives', are defined in More's 'Interpretation Generall' as 'the indivisibility of the inmost essence it self; the pure essentiall form I mean, of plant, beast, or man, yea of angels themselves, good or bad' (p. 422). All things that have being in the scale of creation, angelical to vegetal, are attributed this indivisible, atomic essence. More locates 'pure essentiall' forms deep within living things themselves, even within plant matter. These are the fragments of 'indiscerpible' spirit that long to reconnect with the overarching world soul. His monadic explanation of creation coincides with his vitalist atomism: More is interested not merely in atoms of matter but in atoms of spirit, and it becomes increasingly difficult to distinguish between the two. Book I, Canto 2 of *Psychathanasia*, More's longest poem, concludes with the philosopher's most developed argument for atomism expressed in poetry. He addresses the argument of infinite divisibility, those who claim that God's 'division never could exhaust / The particles ... of quantitie', by insisting that such a theory forgets 'the reverend laws of pietie': 'What thing is hid from that all-seeing light? / What thing not done by his all-potencie?' (1.2.54) For More, the idea that there could be no material indivisibility is contrary to the Christian universe, as it implies that God does not know the 'number, measure, weight' of every particle of His creation (1.2.55). He reasons 'So must sleight Atoms be sole parts of quantitie' (1.2.56), but his concluding note in the final stanza is to acknowledge the 'perplexitie' of the topic in hand:

> 'tis not right
> To reason down the firm subsistencie
> Of things from ignorance of their propertie.
> Therefore not requisite for to determ
> The hid conditions of vitalitie
> Or shrunk or sever'd; onely I'll affirm
> It is, which my next song shall further yet confirm. (1.2.59)

Fallible mortal philosophers cannot know the 'propertie', or 'hid conditions' of the unfathomable units of creation. What More does claim to know, however, is that these atoms are the foundations of 'vitalitie' – life – and that '*it* is' (italics mine.) The subject of 'it' here remains undefined, but appropriately so. More's wording recalls Traherne's later conclusion on the invisible particle, an unfathomable object that, paradoxically, is also fully knowable: something that cannot be 'known by causes' but 'the Matter of it' can be 'felt' by the mind.[29] The author of *Psychathanasia* cannot fathom the properties of a single atom, but he can feel its truth not only through his reason, but with his faith. The 'next song' referenced here may be *Antimonopsychia*, his 'Confutation of the Unity of Souls', which followed on as the third and final poem in the 1642 edition and continued More's investigation into the 'profound and central energie' of individual ensouled beings (p. 294). It could refer equally, however, to the neighbouring canto of *Psychathanasia*, in which More is inspired by the vision of a 'Nymph', Psyche, to continue his song of the soul and of

---

[29] Quoted from Thomas Traherne, *Commentaries*, in *The Works of Thomas Traherne*, ed. Jan Ross, 9 vols (Cambridge, 2005–), III, p. 341. Traherne was inspired by More's writings, quoting from his *Divine Dialogues* at length in his commonplace book. I discuss this influence further in Chapter Two.

immortality (p. 95). Either way, More recognised something inherent to the poetic medium, his 'song', to 'confirm' his theories of the central indivisibility of being.

It is no coincidence that More's most substantial poetic argument for atomism follows a detailed exploration of the indivisible centre of things and what this means for being human. Earlier in the second canto, he asserts

> Forthy let first an inward centre hid
> Be put. That's nought but Natures fancie ti'd
> In closer knot, shut up into the mid
> Of its own self; so our own spirits gride
> With piercing wind in storming Winter tide,
> Contract themselves and shrivell up together,
> Like snake the countrey man in snow espi'd,
> Whose spright was quite shrunk in by nipping weather.
> From whence things come, by fo-man forc'd they backward hither.
>   (1.2.28)

This 'inward centre hid', More proceeds to clarifie, is 'the centre seminall' (1.2.30) common to all living things. Whether plant, animal or human, all beings embody this point of contraction, the 'closer knot' recollective of John Donne's 'subtle knot, which makes us man' in 'The Ecstasy'.[30] As More shows here with the simile of the snake, this 'centre' is susceptible to threat from physical forces, but it offers protection while the body suffers the consequences. It remains constantly 'ti'd / In closer knot' while the corporeal vehicle undergoes change. Arguably, the stanza assumes a parallel function, as it forms a point for the condensing of vital and spiritual information, enfolding and protecting deeper meaning like the 'closer knot[s]' of living things. More's pithy and regular Spenserian verses provide the corpuscular building blocks of his metaphysical reflections. They serve as external reminders of – and as vehicles for reconnecting with – this 'inward centre hid', the 'Atom-life' at the core of all beings.

---

[30] John Donne, *The Complete English Poems,* ed. A. J. Smith, 4th edn (London, 1996), p. 55.

In More's vitalist universe all creatures embody an indivisible essence, whether tree, snake or man. He is clear to distinguish, however, between the restrictions of vegetal life and the superiority of humans. In an especially vivid moment, he acknowledges how plants have a 'soul, whose life in their sweet growth we find' and continues

> So plants spring up, flourish and fade away,
> Not marking their own state: they never found
> Themselves, when first thy 'pear'd in sunny day;
> Nor ever sought themselves, though in the ground
> They search full deep (1.2.15)

This passage is reminiscent of Milton's famous 'One first Matter all', Raphael's lesson to Adam on how

> ... body up to spirit work, in bounds
> Proportioned to each kind. So from the root
> Springs lighter the green stalk, from thence the leaves
> More airy, last the bright consummate flower
> Spirits odorous breathes: flowers and their fruit
> Man's nourishment, by gradual scale sublimed
> To vital spirits aspire, to animal,
> To intellectual, give both life and sense[31]

The echo between More's plants that 'spring up, flourish' and Milton's stalk that 'springs lighter' suggests that the creator of *Paradise Lost* may have been familiar with More's *Philosophical Poems*.[32] Elsewhere, More writes of 'reason down headlong hurld / Out of her throne' by those who would seek knowledge based on sense

---

[31] John Milton, *Paradise Lost*, ed. Alistair Fowler, 2nd edn (Harlow, 2007), Book 5, 478–85.

[32] Jackson Campbell Boswell, in *Milton's Library: A Catalogue of the Remains of John Milton's Library and an Annotated Reconstruction of Milton's Library and Ancillary Readings* (New York, 1975) lists More's works as possible inclusions, citing the critic Robert H. West for further information (pp. 174–75). West acknowledges that Milton 'does not mention Henry More' directly, but the two men were contemporaries at Christ's College and '[a]fterwards More's works were too many and too prominent for Milton to have been oblivious of them'. Robert H. West, *Milton and the Angels* (Athens, 1955), p. 102.

alone – language strikingly similar to Milton's violent fall of Satan, 'Hurled headlong flaming from the ethereal sky / With hideous ruin and combustion down'.[33] The *Philosophicall Poems* contain moments of genuine philosophical beauty in addition to phrasings of terrible clumsiness, such as the clunky 'Long have I swonk with anxious assay' which appears only three stanzas prior to More's resonant plant imagery (1.2.12). There are parallels between More and Raphael's take on vegetal growth, but their respective cycles of generation take off in opposite directions. Contrarily to Milton's account of sublimation on a 'gradual scale', More's plants 'fade away' because 'they never found / Themselves'.

The ability to connect with one's 'inward centre' and thus to 'find oneself' is a key impulse of the philosophical poems. Of More's many neologisms to come from the *Philosophicall Poems*, a large number are 'self' compounds (i.e., 'self-active', 'self-central', 'self-feeling') and others strongly relate to the act of self-reflection (such as 'autaesthesy', translatable as 'self-consciousness').[34] According to *Psychathanasia*, the distinction between humans and other created beings is that the former are able to find their inner 'Atom-lives', and in so doing to mark 'their own state'. Humans, unlike plants, can think beyond the immediate earth of their surroundings and 'aspire' to purer spiritual being. This is because they possess the capacity for reason, though More is quick to point out that this may not be an attribute owned by humans alone. He cites such mammals as 'The Dog, the Horse, the Ape, the Elephant' as intelligent beings who would 'claim their share in use of right reason', then moves on quickly from this possibility as an issue outside his immediate concern, which is the life of the human soul: 'the Image of the highest God, / Which brutes partake not of' (1.2.17–19). In order to find themselves, More explains, beings must not only possess reason but 'reason and reflect / Upon their reasoning' (1.2.18). It is this embedded act of self-reflection that keeps the 'knot' of the 'inward

---

[33] More's quotation is from *Democritus Platonissans* (1646), p. 3; the moment from *Paradise Lost* is 1.45–46.

[34] For an example of 'autaesthesy', see *Psychozoia*, 1.26: 'King of *Autaesthesy* / Is that great Giant who bears mighty sway, / Father of Discord, Falshood, Tyranny'.

centre' firmly tied. The spiritual centres of plants and animals are immortal in the sense that they do not disappear following the creature's corporeal decay, as they are relocated elsewhere in the natural world.[35] But they do not share the immortality of the human soul which, through rational self-knowledge, contracts into an indivisible centre that retains its form despite vehicular change. More confirms this in the next canto:

> In Earth, in Aire, in the vast flowing Plain,
> In that high Region hight Aethereall,
> In every place these Atom-lives remain,
> Even those that cleeped are forms seminall.
> But souls of men by force imaginall
> Easly supply their place, when so they list
> Appear in thickned Aire with shape externall,
> Display their light and form in cloudy mist,
> That much it doth amaze the musing Naturalist. (1.3.28)

The first half of the stanza accords with his previous comments on the endurance of central lives in the world, whether animal or vegetable. In the final five lines, though, he attributes 'souls of men' with the ability to 'supply their place' to the amazement of the 'musing Naturalist'. To 'amaze' could be applied here with the sense of inspiring wonder, which would be suitable, but it also seems to be about perplexity. More uses 'amazen' with this sense in *Democritus Platonissans*, where he considers the impact of obscure visions that 'amazen may / The wisest man and puzzle evermore' (sig. C2v). While the language of physics remains necessary – the Naturalist still views the phenomena amidst a setting of 'thickned Aire' and 'cloudy mist' – the sense in *Psychathanasia* is that the immortal, indivisible 'Atom-lives' of human beings are beyond the enquiries of natural philosophy. The force 'imaginall', the power to transcend or sublimate (as expressed by Milton's Raphael) belongs to metaphysics, and most especially poetry.

---

[35] Speaking more generally, More explains that lives 'their due time do stay, / Though their strong centrall essence never can decay' in the following canto of *Psychathanasia* (1.3.27).

It is therefore entirely appropriate that More begins his philosophical exploration with poetry, the medium best suited to observe the emblematic structure of the universe and to reflect on one's self. Poetry offered equal parts introspection – a glimpse of the 'Atom-life' within – and flights of fancy throughout the macrocosm. More follows the example of the Elizabethans Philip Sidney, George Puttenham and his beloved Spenser in composing poetry that was designed to expound important truths about self, nature and God.[36] He inherited a poetics that was steeped in centuries of Platonic influence and inextricably tied to his own developing philosophy. Kenneth Borris has written of Renaissance poetics:

> In Renaissance writings favouring poetry, certain doctrines and concerns particularly indicate Platonic influence: (1) the poet's divine furor; (2) poetry's uplifting expression of beauty; (3) its 'cosmo-poesis' or symbolic manifestation of the splendid cosmos, deemed God's creation; (4) its legitimacy, according to its instructive ethical, political, epistemological, and theological value for individuals and society; (5) its mimetic relationship to the world's appearances, conceived as reflections of 'higher realities,' and poetry's consequent powers for raising readers' minds towards truth; (6) its properly characteristic allegorism.[37]

---

[36] Philip Sidney writes of how the poet gives 'right honour to the heavenly Maker of that maker, who, having made man to His own likeness, set him beyond and over all the works of that second nature; which in nothing he showeth so much as in poetry, when, with the force of a divine breath, he bringeth things forth surpassing her doings'. He also argues 'that of all writers under the sun the poet is the least liar' as poetry labours not 'to tell you what is or is not, but what should or should not be'. See *The Defence of Poesy* in Sidney's *'The Defence of Poesy' and Selected Renaissance Literary Criticism*, ed. Gavin Alexander (London, 2004), pp. 1–54 (pp. 9–10; p. 35). George Puttenham refers to the 'divine instinct – the Platonics call it *furor*' of the poet at the opening of *The Art of English Poesy*. Quoted from *The Defence of Poesy* in Sidney's *'The Defence of Poesy' and Selected Renaissance Literary Criticism*, pp. 55–203 (p. 57).

[37] Kenneth Borris, *Visionary Spenser and the Poetics of Early Modern Platonism* (Oxford, 2017), p. 37.

It is as though More read a manual on what poetry could do for Platonism and followed it closely. Of the six characteristics of Platonic poetry Borris identifies above, the first (divine furor), the third (cosmo-poesis) and the fifth (reflecting on higher realities) are the strongest foci of More's poetics. He considers the instructive value of his compositions (number four on Borris's list) to be of vital importance, but he struggles with adequacy of expression in delivering didacticism, a sensitivity to the need for precision that results in extensive notes on each of his poems and the accompanying 'Interpretation Generall'. Allegory (number 6) provides the mode for the first of his Songs of the Soul, *Psychozoia*, but his later poems develop a more mimetic exploration of cosmic structures. Beauty presents a more complicated case. More was undeniably attracted to earthly matters, especially as he progressed in poetic composition. While *Psychozoia* attributes its words of philosophical and divine truth to its most beautiful and well-proportioned allegorical characters, the later poems *Psychathanasia* and *Democritus Platonissans* are as concerned with testing hypotheses as reflecting on Pythagorean symmetries. This focus, alongside the undeniable ugliness of occasional turns of phrase ('the crusty fence / Of constipated matter close compresse', *Psychozoia*, 1.28) and the dogged explanatory notes, may have been what led John Hoyles to lament the style of these poems, claiming that More 'felt it necessary to justify his poetry by making it as prosy as possible. He found it difficult to reconcile his lyrical instincts with the demands of philosophical argument.'[38]

But no such reconciliation was necessary, as the 'philosophical argument' of More's first writings poured from the 'lyrical instinct' of poetic inspiration, in an act strongly imitative and affirming of the emanatory cosmic structure explored by the poems. Regardless of tensions in style and execution, it was the *process* of writing poetry that encouraged deeper self-reflection and contact with essential 'energies' – and this 'furor' was strongly associated with a Neoplatonic understanding of the beautiful. More was writing

---

[38] John Hoyles, *The Waning of the Renaissance 1640–1740: Studies in the Thought and Poetry of Henry More, John Norris and Isaac Watts* (The Hague, 1971), p. 69.

prior to the separation of the sublime from the beautiful that took place just over a century later, within a philosophical framework that associated one with the other: awe-inspiring and immense spiritual visions with the perfection of divine beauty. Hence, in the words of Verena Lobsien, 'Neoplatonic aesthetics rests on a structure of excess, and it is hardly separable from its ethical dimension.'[39] Her statement, though directed generally, offers a keen insight into the moral direction of More's excessive poetic outpourings. The overwhelming energy More experienced upon conceiving his poetic visions is well documented, both by the poet himself and by his student and biographer Richard Ward. In 1679, More prefaced his *Opera Philosophica* with an autobiographical statement, in which he reflected that he was 'Stir'd' up to compose 'a pretty full Poem call'd Psychozoia' by 'some Heavenly Impulse of Mind'.[40] As Craig Staudenbaur has neatly put it, More had 'embarked on a course of study and purification' which led to 'a conversion so profound and complete that he felt the need to express it in poetry'.[41] 'Feeling the need' is rather a glib phrase for More's poetic impulse, though an accurate description in literal terms. More explains in the *Opera Philosophica* that he composed the poetry with 'no other Design, than that it should remain by me a private Record of the *Sensations* and *Experiences* of my own *Soul*.[42] This 'private Record' provided necessary textual manifestation of the activity within his soul, a method by which he could preserve a copy of his spiritual experiences for further philosophical reflection.

---

[39] Verena Olejniczak Lobsien, *Transparency and Dissimulation: Configurations of Neoplatonism in Early Modern English Literature* (Berlin, 2010), p. 11.
[40] More, Preface to the *Opera Philosophica*, trans. Richard Ward. Quoted from Richard Ward, *The Life of Henry More by Richard Ward*, ed. Sarah Hutton, Cecil Courtney, Michelle Courtney, Robert Crocker and Rupert Hall (Dordrecht, 2000), p. 21.
[41] Craig Staudenbaur, 'Galileo, Ficino, and Henry More's *Psychathanasia*', *Journal of the History of Ideas*, 29.4 (1968), 565–78 (569–70).
[42] More, Preface to the *Opera Philosophica*, quoted from *The Life of Henry More by Richard Ward*, p. 21.

More did not intend his first working of the poems for publication, and it was under pressure from colleagues that *Psychodia Platonica: or a Platonicall Song of the Soul* appeared from Cambridge University Press in 1642. When reworking the poems into an improved, expanded collection in 1647, he was conscious to address what he considered the imperfectly formed compositions of the earlier publication, explaining his creative process in an epistle 'To the Reader, Upon this second Edition':

> I have taken the pains to peruse these Poems of the soul, and to lick them into some more tolerable form and smoothnesse. For I must confesse such was the present haste & heat that I was then hurried in (dispatching them in fewer moneths then some cold-pated-Gentlemen have conceited me to have spent years about them, and letting them slip from me so suddenly, while I was so immerse in the inward sense and representation of things, that it was necessary to forget the oeconomie of words, and leave them behind me aloft, to float and run together at randome (like chaff and straws on the surface of the water). (sigs Br–v)

More responds here to a long literary tradition of addressing one's poetry as one's offspring, but with a twist. His reference is not to express anxiety or pride in the progeny let loose in the world, but his feelings at an earlier stage of the upbringing – on poems taking form at the hands, or indeed tongue, of the parent. Following Pliny, early modern natural histories spread the opinion, in the words of Thomas Browne (who would proceed to dismiss the 'vulgar' claim) that 'a Bear brings forth her young informous and unshapen, which she fashioneth after by licking them over'.[43] More returns to fashion his young stanzas late, with the apology that he had released them too soon in what was 'the present haste & heat' of his inspiration, an appropriate defence of his excess in creation following his 'Heavenly Impulse'. He then moves on to another striking metaphor. 'Oeconomie' is likely in reference to the general management of his words; it could also respond to the need of building a structure between his ideas and externalised observations, which, although

---

[43] Thomas Browne, *Pseudodoxia epidemica: or, Enquiries into very many received tenents, and commonly presumed truths* (London, 1646), Book 3, Chapter 6, p. 116.

necessarily interdependent as stemming from the one inspired mind, were left in their 'informous' state 'to float and run together at randome'. More's simile for clarification here is 'chaff and straws' strewn on water, but it could so easily have been atoms. The imagery of fragmentary forms and 'randome' collisions smacks of corpuscular activity from an author known to favour atomism.

More attributed atomic construction to the shaping of his poems, just as he did to the working of the divine cosmos. In another epistle from the 1647 volume, 'To his dear Father Alexander More Esquire', he wrote self-deprecatingly 'I am not indeed much solicitous, how every particle of these Poems may please you', admitting nonetheless 'I please my self in the main; which is, The embalming of his name to Immortality, that next under God, is the Author of my Life and Being' (sig. A3r). His wording again associates a vitalist interpretation of matter, here the understanding that his text will preserve the name of his father, with an atomic structure that develops poetry particle by particle. Not just a passing reference, More's address taps into a key understanding of the nature of his universe and the expectations of poetic imitation. In the preface to *Democritus Platonissans* he presents his argument for a plenum, explaining that 'if any space be left out unstuff'd with Atoms, it will hazard the dissipation of the whole frame of Nature into disjoynted dust' (sig. A2r). More follows Descartes here, with the exception that his particles are true indivisibles rather than Cartesian corpuscles. If his universe is a plenum, so are his poetic compositions: the stanzas of *Psychozoia*, *Psychathanasia* and *Democritus Platonissans* may feel as 'stuff'd' as they do from an attempt to combine their particles and support 'the whole frame' of More's excessive Neoplatonic vision. An atomic structure, a world built from indivisibles within the indivisible God, reveals a Neo-Pythagorean vision of the many in the one. So the reader encounters a typically replete stanza from *Psychozoia*, brimming with content as the poet introduces the first of the four 'films' of Psyche, Physis (interpreted by More in his notes as 'Nature vegetative'):

The first of these fair films, we Physis name.
Nothing in Nature did you ever spy,
But there's pourtraid: all beasts both wild and tame,

> Each bird is here, and every buzzing fly;
> All forrest-work is in this tapestry:
> The Oke, the Holm, the Ash, the Aspine tree,
> The lonesome Buzzard, th'Eagle and the Py,
> The Buck, the Bear, the Boar, the Hare, the Bee,
> The Brize, the black-arm'd Clock, the Gnat, the Butterflie. (1.41)

As a stanza, this initial description of Physis strongly suggests the poet's concern that nothing be 'left out unstuffd'. Interestingly, it diverts from More's usual Spenserian rhyme-scheme, with the stanza concluding CBCB rather than CBCC, as customary. This feels purposeful, even by the rough standards of More's imitation of Spenser. One stanza was not enough to offer a full, fitting description of what Physis has to offer. The following verse continues:

> Snakes, Adders, Hydrae, Dragons, Toads and Frogs,
> Th' own-litter loving-Ape, the Worm, the Snail,
> Th'undaunted Lion, Horses, Men, and Dogs,
> Their number's infinite... (1.42)

More's 'stuffd' stanzas correspond directly with the 'inward centre' of all living things, the 'Natures fancie ti'd / In closer knot' which, due to their superiority of reasoning, human beings are equipped to find through the process of self-reflection. As the central line of stanza 41 declares, 'All forrest-work is in this tapestry' of physical life: the poem, like the 'Atom-life', holds the source of all vitality. More's Spenserian stanzas function both as sites of excess and as points of contraction. In content, they are like the vehicles of 'vital congruity', brimming over with substance; in structure, they complement the 'closer knots' of More's indivisible units as they congregate to build their points. Typically, the middle lines of each verse present its main, driving argument, following an opening that establishes the subject and preceding a close that connects the discovery with the topic of the next stanza.[44] Through the act of writing or reading these particulate stanzas, the Christian subject

---

[44] This presents a contrast with Spenser's stanzas in *The Faerie Queene*, which typically present their argument in the alexandrine. I am grateful to Liza Blake for this observation, 24 June 2020.

can 'reason and reflect / Upon their reasoning', thus increasing self-discovery of the indivisible essence within.

The 'stuffd' plenum of More's verse owes as much to the influence of Spenser as it does to his quasi-physical metaphysics. More's prefatory letter to his father acknowledges his debt to the parent who lays a 'more proper claim' to the poems 'then any', having

> from my childhood turned mine ears to Spencers rhymes, entertaining us on winter nights, with that incomparable Peice of his, The Fairy Queen, a Poem as richly fraught with divine Morality as Phansy. (sig. A2v)

Alexander More introduced his son to Spenser, and the growing philosopher recognised the Platonic beauty of his bedtime stories. The significance of More's adoption of the Spenserian stanza for his poetry cannot be understated. Firstly, akin to Lobsien's assessment of the ideal Neoplatonic aesthetics, *The Faerie Queene* to More's mind similarly 'rests on a structure of excess', and its creative value is 'hardly separable from its ethical dimension'.[45] The 'richly fraught' content describes the poem as a vessel laden with Platonic virtue, poured from an enraptured 'Phancy' that is enabled by and revelatory of Christian morality. Secondly, 'Spencers rhymes' get a special mention. More claims that his ears were 'turned' to the regular rhymes and rhythms of *The Faerie Queene*'s verses. It was not just childhood sentimentality that led the Neoplatonist to Spenser, but an awareness of immanent spiritual productivity in the Elizabethan poet's choice of form. Like John Davies of Hereford before him – whom, perhaps pointedly, he does not acknowledge – More lays out his *Platonick Song of the Soul* entirely in Spenserian stanzas.[46]

The constancy of the Spenserian form develops a sense of infinity in the structure that mirrors the poetic content. Jeff Dolven suggests that Spenser's stanzas 'must have come to serve him [Spenser] as a kind of architecture for thinking, both in a reliable framework and a constant provocation, installing in every unit of verse both

---

[45] Lobsien, p. 10.
[46] John Davies of Hereford, *Microcosmos: The discovery of the little world, with the government thereof* (London, 1603).

a second thought and an impulse to closure.[47] More's 'particles', similarly, stand individually but resist satisfying the 'impulse to closure', urging the reader to take another step in a never-ending exploration of the universe and the self. The corpuscular 'framework' of More's Spenserian stanzas combines with the atoms of his plenum, resulting in a vital poetics and what Lobsien defines as a Neoplatonic resistance to 'discursive completeness'.[48] More's poetic spaces, powered by vitality and excess, create a plenum parallel to his Neoplatonic cosmos wherein all things connect and project meaning. His stanzas behave like connecting particles, joining together to form the composite, comprehensive structures of both philosophical content and lengthy poem. Each stanza is also, however, a plenum on its own, overflowing as a 'receptacle' for the essence of vitality. As 'pure essentiall' forms, 'Atom-lives' accommodate both the indivisible source and its consequences, the continuous activity of all things pouring through creation.

More's atoms are not the 'dusty Atoms of Democritus', which he condemns at the opening of *Psychathanasia*. While the particles of Democritus and Epicurus supported a philosophy that 'souls all mortal be', his indivisibles are an inseparable part of internal, immortal life. Yet in 1646 he chose to publish a poem called *Democritus Platonissans* – a bold move from a confirmed Neoplatonist who had rejected the 'putid muse' of Lucretius just four years earlier.[49] More was attracted to the Democritean heritage of indivisible atomism which, as a principle he traced back to Pythagoras, was entirely reconcilable with his Neoplatonic understanding of the cosmos. In writing a long philosophical poem on the nature of being, albeit from a Christian–metaphysical perspective rather than the Epicurean, More must have been conscious of his great

---

[47] Jeff Dolven, 'Spenser's Metrics', in *The Oxford Handbook of Edmund Spenser*, ed. Richard A. McCabe (Oxford, 2010), pp. 385–402 (p. 392).

[48] Lobsien, p. 86. Lobsien's argument here is in direct reference to a 'double suspension of certainty' in Andrew Marvell's Platonic poetry, which she claims replaces 'static representation with complex interaction'. This theory can be applied productively to More's vital poetics.

[49] More first rejected the 'putid muse' in *Psychathanasia*, 1.1.6, *Psychodia Platonica*, sig. E6v.

Latin predecessor. Alexander Jacob argues that More's 'Platonick Song' was intended as 'Anti-Lucretian Poetry', which aptly summarises his stance: while Lucretius, Jacob points out, 'rarely looks "behind" things', More's subject 'is the spiritual basis of the continuity of consciousness, and his entire work may be read as an attempt to chart the persistence of memory through all forms of life'.[50] 'All forms of life', in this context, include the very atoms of this vitalist, ensouled universe.

It is with this work, More's most explicit poetic experiment with Neoplatonic atomism, that I wish to conclude this chapter. The poetry of *Democritus Platonissans* is direct in displaying the differences between More's atomism and the Democritean inheritance marked in the title. The verse draws on a typical Morean word – 'conspissate', that which is thickened, or made dense – to consider the vitality of enlivened matter:

> And that which conspissate active is;
> Wherefore not matter but some living sprite
> Of nimble nature which this lower mist
> And immense field of Atoms doth excite,
> And wake into such life as best doth fit
> With his own self. As we change phantasies
> The essence of our soul not chang'd a whit
> So do these Atomes change their energies,
> Themselves unchang'd into new Centreities. (Stanza 14, sig. B2v)

At this point in the poem More's material philosophy takes a startling turn – it appears to deny that there is such thing as a state of matter distinguishable from spirit. The distinction he makes elsewhere between material and immaterial states of being is notoriously cloudy, as discussed earlier, but the poetic space of his Platonic–Democritean 'hypothesis' poses an emboldened view on the relationship. The previous stanza concluded in a strong, Neoplatonic vein, with More characteristically adhering to Plotinus's view that 'body's but this spirit [Psyche], fixt, grosse by conspissation'. This on its own does not deny

[50] Alexander Jacob, 'The *Platonick Song* as Anti-Lucretian Poetry', in Henry More, *A Platonick Song of the Soul: Henry More*, ed. Alexander Jacob (Lewisburg, 1998), pp. 16–19 (p. 18).

the polarisation of matter from spirit, as the fragments of the Plotinian 'life spark' are so fallen as to become unrecognisable in substance and form. But in the following, above-quoted stanza, he informs us that 'conspissate' substance is not so 'fixt' after all: it is 'active'. That which is 'conspissate', More informs us, is inspired by 'some living sprite of nimble nature' which excites the 'immense field of Atoms'. His atoms here are strongly associated with vitalistic, vegetative growth; a crop watered by some semi-naturalistic, semi-supernatural 'living sprite' that, like one working the land, harvests from the seeds what is needed for the season ('as best doth fit / With his own self'). The indirect but clear comparison between atoms and seeds is strongly Lucretian, but the accompanying psychology is not.[51] Rather, More raises the state of matter closer to that which is immaterial and immortal, a state of being emphatically denied by Epicureanism. He embellishes his point about the activation of 'conspissate' matter with an analogy that recalls the opening of the poem. 'As we change phantasies', he reminds us, the 'essence of our soul [is] not chang'd a whit'; each individual, ensouled human being has their own particular thoughts and 'phantasies', but the essence of their spiritual life-force remains the same. The self-reflective acts of his poetry frequently identify changing interactions and restlessness within the soul, but the second stanza of *Democritus Platonissans* explains:

> Strange sights do straggle in my restlesse thoughts,
> And lively forms with orient colours clad
> Walk in my boundlesse mind, as men ybrought
> Into some spacious room, who when they've had
> A turn or two, go out, though unbad.
> All these I see and know, but entertain
> None to my friend, but who's most sober sad;
> Although, the time my roof doth them contain
> Their presence doth possesse me till they out again. (sig. Br)

---

[51] For an example of the Lucretian association between atoms and seeds see the opening of *De Rerum Natura*, in the words of Lucy Hutchinson's translation: 'What we in reasoning the first matter call / Generative bodies, and the seeds of all'. See *The Works of Lucy Hutchinson, Vol. 1: The Translation of Lucretius*, ed. Reid Barbour, David Norbrook and Maria Zerbino (Oxford, 2012), Book 1, lines 51–52 (p. 21).

At the beginning of the poem More denounces the futility of his senses one by one ('Hence, hence unhallowed ears…'; 'foul tongue I thee discard…' 'Strange sights do straggle…', stanzas 1–2, sig. Br). He acknowledges, as he does in the context of stanza 14, that the possession of 'restlesse thoughts' brings only temporary change to the 'spacious room' of his soul. The present-day reader could be reminded of the emptiness of T. S. Eliot's refrain 'In the room the women come and go / Talking of Michelangelo' from 'The Love Song of J. Alfred Prufrock'.[52] But while the coming-and-going of Eliot's famous poem denotes modernist hollowness, More's 'spacious room' is designed to be filled with the divine and is only momentarily diverted by fragmentary things. His 'friend' is his 'pregnant Muse', his point of contact with the divine who will acknowledge no thought 'Till Eagle-like she turn them to the sight / Of the eternall Word all deckt with glory bright' (stanza 1, sig. Br).[53]

Seemingly, More is comparing the 'living sprite / Of nimble nature', which infuses the atoms with life, with the human soul or mind that activates, entertains and is temporarily possessed by shifting thoughts, but remains essentially unchanged. The 'phantasies' analogy of stanza 14, however, is not directed at the 'sprite' but at the atoms themselves, which have been woken by it. It is not Neoplatonic plastic spirit that More is associating with the unchanging human soul, but atoms. These 'Atomes', like human minds, 'change their energies' but their essential form – the Morean word 'centreity' – remains constant. Pointedly, his work most inspired by Lucretius and the 'mechanical' philosophy of Descartes picks up on a common Lucretian motif, the similitude between human beings and atoms, as a method of Platonising the atom and distancing it from its Epicurean origins. More's atoms are no senseless particles, but matter ('conspissate' Psyche) infused with spirit, activated and subject to different 'energies' just like individual men and women. As the reader completes stanza 14 and discovers the

---

[52] 'The Love Song of J. Alfred Prufrock', in T. S. Eliot, *Selected Poems* (London, 1961), pp. 11–16 (p. 11).

[53] More returns here to a common Neoplatonic image, that of the eagle to represent soul (I have discussed his use of the image in *Psychathanasia*, 2.1.10; see pp. 47–48 of this chapter.)

unexpected twist of the analogy, the subject of the singular possessive pronoun in the central lines becomes unclear. The 'living sprite' wakes each atom 'into such life as best doth fit / With his own self'. This raises the question – is the 'self' in control of the life a guiding spirit, or the constant 'Centreity' of each individual atom? More's atoms do more than function as metaphorical models for what it means to be human: they constitute what they represent. *De Rerum Natura*, as William Shearin has recently argued, homogenises human beings through the repeated personification of atoms, and vice versa, as atoms lack distinguishing characteristics and are all one and the same, *unus quisque*.[54] From this argument we get the concept of atomisation, again recalling the empty horror of the faceless masses often associated with modernist poetry. An atomised population has lost all traces of individualism and is lacking in social bonds, resulting in the disintegration of relationships parallel to the breaking down of material compounds. While there is a sense of fragmentary loss to More's world-picture – the forceful separation of individual souls from their composite origin – no item of his Neoplatonic universe is without its correspondence with all other existing things in the scale of creation. The similitude between people and atoms in *Democritus Platonissans* is not aimed at human bodies, but at souls – entities which, like atoms, are in 'essence' all one and the same. These stem from and return to the same source, essentially unchangeable even as they flood with individual thoughts and activities. More continues:

> And as our soul's not superficially
> Colour'd by phantasms, nor doth them reflect
> As doth a looking-glasse such imag'rie
> As it to the beholder doth detect:
> No more are these lightly or smear'd or deckt
> With form or motion which in them we see,
> But from their inmost Centre they project
> Their vitall rayes, not merely passive be,
> But by occasion wak'd, rouze up themselves on high. (stanza 15, sig. B2v)

---

[54] W. H. Shearin, *The Language of Atoms: Performativity and Politics in Lucretius' De Rerum Natura* (Oxford, 2015), pp. 141–63 (pp. 142–43).

The essential word here is 'vitall'. More's atoms are not externally dressed or 'smear'd' with attributes, but project energy from their 'inmost Centre'. Stanza 16 continues the analogy between atoms and people, to the extent that the reader could forget that the comparison was introduced as a simile in the first place. Like humans, these atoms are ensouled, as More demonstrates through a classic Neoplatonic image of projection. They are particles of a vitalist universe, endowed with self-action, the necessity of which More clarifies in the following verse: 'they're life, form, sprite, not matter pure, / For matter pure is a pure nullitie, / What nought can act is nothing, I am sure' (sig. B3r). The final line of stanza 15 nevertheless takes this emphasis on ensouled, active matter to a surprising place. More argues that atoms 'by occasion wak'd, rouse up themselves on high' (sig. B2v). It is difficult at this point to distinguish between what is intended as metaphor and what corresponds with the literal foundations of More's entwined physics and metaphysics. Physically, the image of awakening and rising fits with the motions of the Platonised Democritean atom, which remains constant even as it moves from place to place across the vast – and, according to this thought experiment, potentially infinite – universe. An atom in the mud one day could be in the sun the next, as Traherne would go on to explore in his atomic poetry two decades later.[55] Yet the emphasis on 'vitall rayes', awakening and rising 'high', cannot but evoke the rising of the soul in its final stages of vital congruity. Atoms, like embodied human souls, are 'wak'd' and 'find themselves', thus supporting the return of spirit to the divine source.[56]

Atoms assume a parallel status to embodied human souls. They are more than merely body, *Democritus Platonissans* tells us, because we cannot conceive of atoms in motion – in other words, atoms *as* atoms – unless they possess a vital force from ensoulment. Earlier in this chapter, I raised the question of how More's early atomism related to his theory of vital congruity. The answer is that his poetic works establish an intimate relationship between the two. The vehicle of poetry facilitated More to draw an emblem of

---

[55] *Commentaries*, in *Works*, III, p. 349. See also Chapter Two, p. 105.
[56] Stephen Clucas also acknowledges More's active, 'living atoms' in 'Poetic Atomism', 338.

the ensouled, 'vitall' atom, which in turn provided a vehicle to aid understanding of the ambiguities around the embodied soul. At a minute scale, the atom assumes a parallel existence to the earth-drawn soul undergoing the stages of vital congruity. This relationship extends beyond metaphor, as the emanationist structures of More's cosmos depend upon these projections and parallels of motion and matter infused with spirit. The sketching of atoms, the founding 'knots' on Physis's stole, as 'life, form, sprite, not matter pure' explains the necessity of the ambiguities surrounding vital congruity – there can be no straight severance between matter and soul in a cosmos where all things are permeated by the deity. The 'knots' of Physis's stole are dispersed throughout the cosmos to form material things, while the inner 'Atom-lives' of human beings remain firmly tied and connected to their divine source.

If, as *Democritus Platonissans* implies, lowly atoms can find themselves, so must human beings. The phrase 'find yourself' might smack of the twenty-first-century self-help book, but the act of recognising and meditating on your innermost core was of vital importance to More's theology. More's thoughts on atomism changed over the course of the following decades, but the Platonised atoms of his *Philosophicall Poems* served a special, spiritual purpose. The atomism of his philosophical poetry provided the means by which self-reflection, or introspection, on the 'Atom-life' of one's own soul was made possible. More's poetic inspiration led him into the indivisible essences of things. As a result, the atom, or monad, became an explorable emblem of the divine.

Chapter 2

THOMAS TRAHERNE'S ATOMS, SOULS AND POEMS

> For Souls are Atoms too, and simple ones.
> Nay less then Atoms, not so firm as Stones,
> They are divested not of Bulk and Size
> Alone, but of all Matter too, and are
> So void, that very Nothings they appear.
> Yet for these Souls all Atoms were prepard[1]

Traherne writes these lines in his *Commentaries of Heaven*, at the end of a long, comprehensive study of the 'Atom'. In this passage, from the second of two poems in the entry, the fine detail of his meditation on the topic produces a bold conclusion: that 'Souls *are* Atoms... / Nay less than Atoms' (italics mine). The focus of his preceding section was on the definition of an 'Atom', its 'Etymologie' and 'Nature', but the closing verse moves the atom from subject to object, with the rational soul taking its place as the dominant (yet, paradoxically lesser) subject.[2] At this point, Traherne reveals the higher purpose of considering the atom by

---

[1] *The Works of Thomas Traherne*, ed. Jan Ross, 9 vols (Cambridge, 2005–), III, p. 362.
[2] *Works*, ed. Ross, III, p. 333. For the 'Atom' commentary, see ibid., III, pp. 333–63. The *Commentaries of Heaven* are the second and third volumes of Jan Ross's *Works*. All further quotations from Traherne's *Commentaries* will be taken from this edition and will be referenced in the text by volume and page numbers.

implying that, through understanding of the one, we come to the other: grasping the atom leads to deeper knowledge of the soul.

Traherne is fond of wordplay. His souls, which have been equated with atoms, are 'simple ones' in the sense that they are utterly devoid of matter, but also because they are capacious, pure first principles, entirely 'at one' with the divine. In the hierarchy of this spiritual relationship, less is more. Souls 'are Atoms' in the sense that the definition of the latter – as indivisible entities – is also applicable to the rational soul. They are 'less then Atoms', however, because they are not only incorporeal but entirely immaterial. Earlier in the commentary, Traherne writes that the soul is 'void' of itself 'that it might be full of All Things' (III, p. 351): the empty, capacious spiritual being has the energy and space to pry into the mysteries of 'All Things', natural and divine. He continues the paradoxical association between material 'Nothing' and spiritual totality in his poem with another double meaning, emerging from the line 'So void, that very Nothings they appear'. From one reading the line argues that souls are like 'Nothings', but from another it claims that 'they appear' *from* 'Nothings'. The syntactical structure has been shifted to place the verb 'appear' at the end of the line and sentence. While concluding with 'Nothings' would have stressed the ineffability of the soul to material experience, the weight on 'appear' marks the presence and positivity of spiritual life. The resulting emphasis is that the soul is accessible to human experience and understanding, in spite or even because of its presence as a *no-thing*, and that this understanding can be achieved by meditating on the 'Atoms' that prepare its activity.

It was in the *Commentaries of Heaven* that Traherne developed his most comprehensive atomic theory.[3] Presented in folio and

---

[3] Even by early modern standards of damage by fire and flood, the *Commentaries* overcame significant obstacles to reach the present-day reader. In 1967 the manuscript very nearly met its end on a burning rubbish tip in Lancashire – thankfully it attracted the notice of a passer-by, Lawrence Wookey, who removed it before damage had spread from the front and back boards to the inner pages. The *Commentaries* then emigrated with their rescuer to Toronto and it was not until 1981, following a chance conversation between the Wookeys and their lodger Elliot Rose, a graduate

written on expensive paper, the *Commentaries* was a voraciously ambitious work. It was an encyclopaedic attempt to investigate 'ALL THINGS ... Created and Increated' in an alphabetically ordered, sequential study of individual subjects (II, p. 3). Here and elsewhere, Traherne stresses the necessity of curiosity and investigation into all subjects, public and private, physical and spiritual, near and far. Even the most unreachable entities of his work, including the 'Increated' deity – that is, the uncreated and everlasting divine being – are sought by the understanding for verbal communication.[4] Traherne equates the pursuit of knowledge with the selfless holy life, so that mankind is obliged to develop the understanding into infinity. He declares in the *Centuries of Meditation*, for example, that it is 'the Nobility of Mans Soul that He is Insatiable' in the pursuit of knowledge.[5] Perhaps as a result, the *Commentaries* text runs over 300,000 words (though sadly it only fulfils its ambitions between the headings 'Abhorrence' and 'Bastard'). Each 'commentary' follows a similar pattern: a definition of the object under consideration, accompanied by critical observations and in most cases a concluding poetic reflection.

Of all the items, the section on the 'Atom' is by far the longest in the text. There was something inherent in the subject of atomic thought that fuelled Traherne's pen, leading him to close with two

---

student at the University of Toronto, that the manuscript was identified as the work of Thomas Traherne. For more information on the manuscript, including its remarkable provenance, see Elliot Rose, 'A New Traherne Manuscript', *The Times Literary Supplement*, 19 March 1982, 324 and Allan Pritchard, 'Traherne's *Commentaries of Heaven* (with Selections from the Manuscript)', *University of Toronto Quarterly*, 53.1 (1983), 1–35 (1). The manuscript is now British Library, Add. MS 63054.

[4] In the posthumously published *Christian Ethicks* (1675), he dedicated a chapter to the 'necessity, excellency and use' of acquiring knowledge and strongly concluded '*TRUE Contentment is the full satisfaction of a Knowing Mind*'. This was followed by a direct attack against the tradition of negative theological consideration: 'That Negative Contentment, which past of Old for so great a Vertue, is not at all conducive to Felicity, but is a real Vice: for to be Content without cause, is to sit down in our Imperfection.' Thomas Traherne, *Christian Ethicks*, ed. Carol L. Marks (Ithaca, 1968), p. 217.

[5] *Thomas Traherne: Centuries, Poems, and Thanksgivings*, ed. H. M. Margoliouth, 2 vols (Oxford, 1958), I, First Century, aphorism 22.

of the longest poems in the *Commentaries* and the editorial query of whether 'it be not best leav out some of these Poems' (III, p. 363). His atoms are not the sterile particles derided by scholasticism, nor the warring elements of Lucretius – they are emblems of Christian eternity and love. In the *Commentaries* Traherne asserts the divine characteristics of the indivisible particle:

> An Atom is like the Heavens, a certain kind of retired Quintessence, far removed from all Elementary Qualities. (III, p. 334)

'Quintessence' is the Aristotelian fifth essence, that which lies beyond the four elements of earth, water, air and fire (and therefore 'far removed from all Elementary Qualities.') This mystical substance was considered to be 'latent in all things; the pure, essential part of any substance' and moreover the 'ultimate substance of which the heavenly bodies were supposed to be composed.'[6] An atom, therefore, is more than either a material particle or an accommodating metaphor: it bears inherently divine connections; it is simultaneously both within and beyond all 'elementary' things. Elsewhere in his writings Traherne is drawn to the same vocabulary when attempting to describe the power of love, which he labels the 'Pure Quintessence of Heaven!'[7] Far from senseless, his atoms are inspired by divine energy and spirit. They may cautiously only be *like* 'the Heavens', but their 'retired Quintessence' is the stuff of God's love – the essential, divine energy connecting all things. It is thus that Traherne concludes a chapter on atoms in his late study *The Kingdom of God*, with the realisation that 'Heaven and Earth dependeth upon their Existence. So do our very Bodies and Lives, and Senses. They are more necessary to us, then the Existence of Angels' (I, p. 347).

---

[6] I quote from Jan Ross's notes on the meaning of 'Quintessence', from *The Works of Thomas Traherne*, I, p. 536.

[7] This exclamation is taken from *The Kingdom of God*, in *Works*, ed. Ross, I, p. 306. All further quotations from this text will be quoted from this edition and volume and will be referenced in the text by page numbers. *The Kingdom of God* is the longest work from the Lambeth Palace Manuscript, discovered by Jeremy Maule in 1997. See Denise Inge and Calum Macfarlane, 'Seeds of Eternity: A New Traherne Manuscript', *The Times Literary Supplement*, 2 June 2000, 14. The manuscript is Lambeth Palace Library, MS 1360.

Traherne was fascinated with the atom as a connecting-point between body and soul, time and eternity, humanity and the divine. His understanding of atomic philosophy was heavily influenced by Henry More, whom he referenced multiple times across his encyclopaedic works. He also copied at length from *Divine Dialogues* in his commonplace book, where More's reflections on cosmic physics do not enter the sections on 'Atoms' or 'Matter', as might be expected, but substantially contribute to Traherne's reflections on the nature of the 'Deitie'.[8] While critics have acknowledged the influence of Cambridge Platonism on Traherne's work, there has been no exploration of his considerable debt to More's atomic theory.[9] Like More before him, Traherne understood the atom as a founding particle not only of material but of metaphysical being. His atoms, like those of More, serve as contractions of divinity in an ensouled material world. Both theologians interpret the atom as an emblem of Christian constancy and Traherne similarly emphasised self-reflection, or the importance of knowing the indivisible core of one's self, as a method for living a divine existence. There are, however, differences between the two writers' respective understandings of the indivisible. Traherne's atoms are associated more closely with human experience. There is a hint of this when he qualifies that they are like 'retired' quintessence, as he argues strongly elsewhere that through 'Retirement [from worldly things] alone can a Man approach to that which is Infinit and Eternal'.[10] Atoms

---

[8] The section on 'Deitie' in Traherne's commonplace book is Dobell Folio, Bodleian Library, MS Eng. poet. c. 42, fol. 33r–v. Elsewhere in the commonplace book, Traherne's reflections on 'Matter' (65v) are taken from Theophilus Gale, *The court of the gentiles, or, A discourse touching the original of human literature, both philologie and philosophie, from the Scriptures and Jewish church* (Oxford, 1670). Traherne quotes from Gale's section on 'Of Plato's Physicks, and their Traduction from Sacred Storie', Book 3, Chapter 9, pp. 312–52.

[9] An exception is Stephen Clucas, who, though he does not suggest a direct influence between the two, discusses More and Traherne together as seventeenth-century literary proponents of the atom in his article 'Poetic Atomism in Seventeenth-century England: Henry More, Thomas Traherne and the "Scientific Imagination"', *Renaissance Studies*, 5.3 (1991), 327–40.

[10] Thomas Traherne, *Inducements to Retiredness*, in *Works*, I, pp. 4–48 (p. 5).

are emblems of Christian retirement, humility and endurance; reminders of the innate yet mysterious bond between base matter and immortal spirit that constitutes the human condition. He adopted More's stance of the metaphysical and poetic atom but made more explicit the benefits of its spiritual content for the individual, flawed human being. Moreover, Traherne moved beyond the Cambridge thinker's emphasis on self-knowledge to contend that an intuitive understanding of the atom is akin to knowledge of soul and God. While self-reflection is important in Traherne's theology, so is the soul's insatiability – its need to connect with and absorb, as he declares on the frontispiece to the *Commentaries*, 'ALL THINGS'. The atom aids it on its quest. If More's atoms are an essential part of the world's metaphysical fabric, Traherne's indivisible particles are the key to restoring contact between the rational soul and the divine. Therefore, he could state unequivocally that the structures of Heaven and Earth rested on the continual act of atoms melding the material world with indivisibility and immortality. Their impact on those created – physically, of course, from atoms – in God's image was far greater than that of the solely spiritual essences of angels.

Such stances mark the writer out as an original, and not a little daring, thinker. Yet Thomas Traherne – Anglican clergyman, devotional poet and exemplary early modern polymath – remains little read even by those with interests in seventeenth-century theology and literature.[11] Meanwhile the poems of his near-contemporaries

[11] Traherne's theological theories were open to as many scholarly influences as his philosophical principles, though it is usually impossible to distinguish between the two. He opens section XVI of the Lambeth Palace manuscript work *A Sober View of Dr Twisses his Considerations*, a discourse on the arguments between the theologians Henry Hammond and Robert Sanderson over the Calvinist writings of William Twisse, with the adamant statement: 'We Study Truth and not Parties'. See *Works*, I, p. 132. Though elsewhere Traherne writes particular attacks against Catholicism, Anabaptism and Socinianism, his Protestantism remains intellectually untied and is influenced by Hammond's Arminianism as well as by Sanderson's Calvinist scholarship (though he comes to favour the latter). For his anti-Catholicism, see the only work he had published during his lifetime, *Roman forgeries, or, A true account of false records discovering*

George Herbert and Henry Vaughan, to whom he is often compared, are privileged by critics, anthologies of early modern poetry and undergraduate reading lists alike. From the very first stages of his modern 'rediscovery' (following the earliest publication of his poetry by Bertram Dobell in 1903), the difficulty of tying Traherne's ideas to any one school of thought has resulted in a long-standing critical misunderstanding that has discredited his achievements.[12] His style is often conceived as outdated. A. L. Clements opens his study of Traherne's mysticism with the claim that 'though contemporary movements surely had some effect on him', the context for his study should be 'the rich and complex Christian contemplative or mystical tradition' of late antiquity and medievalism.[13] More recently, Susan Stewart argued that his poetry in particular demonstrates 'the synonymity of aspects of seventeenth-century lyric with aspects of the medieval'.[14] From this lasting interpretation emerges a claim that Traherne had little interest in or understanding of contemporary scholarly progress. He was not, Richard Douglas Jordan suggests, 'a modern man'.[15] Writing on Traherne's *Centuries*, Carol L. Marks considers his presentation of the 'physical world' as a substance that merits 'study (such as the Royal Society supported and Traherne may have dabbled in)', implying that his interest in natural knowledge was casual at the very least; indeed, she argues that her suggestion is based on the 'slender evidence' of his philosophical imagery.[16] Though the natural philosophical references in the *Centuries* are not as 'slender' as she claims, Marks was writing

---

    the impostures and counterfeit antiquities of the Church of Rome (London, 1673). The *Commentaries* includes a rant against the Anabaptists ('Baptism', III, p. 452) and the Socinians ('Atonement', III, p. 372).

[12] Bertram Dobell published the poems he discovered in *The Poetical Works of Thomas Traherne, B.D. (1636?-1674), Now First Published from the Original Manuscripts*, ed. Bertram Dobell (London, 1903).

[13] A. L. Clements, *The Mystical Poetry of Thomas Traherne* (Cambridge, MA, 1969), p. 13.

[14] Susan Stewart, *Poetry and the Fate of the Senses* (Chicago, 2002), p. 242.

[15] Richard Douglas Jordan, *The Temple of Eternity: Thomas Traherne's Philosophy of Time* (Port Washington, 1972), p. 55.

[16] Carol L. Marks, 'Thomas Traherne and Cambridge Platonism', *PMLA*, 7 (1966), 521-34 (527).

before the discovery of the later Traherne manuscripts, in which the writer's scientific leanings are most prominent. However, more recently in a paper discussing the *Commentaries*, Finn Fordham argued that the clergyman's writings 'frequently show a weak absorption of scientific discoveries'.[17]

The evidence for Traherne's reading in natural philosophy in fact points to quite the opposite of 'weak absorption'. In recent years, studies by James Balakier and Chad Michael Rimmer have begun to explore the complexities of his natural philosophical thought.[18] Traherne's notebooks contain thorough notes on works by More, Robert Boyle and Ficino, amongst many others; he also explicitly names the following philosophers as inspirational thinkers in *The Kingdom of God*: 'Copernicus ... Descartes, Gassendus ... Dr Charleton, Dr Willis, the learned Gale ... my Lord Bacon, Sir Kenelm Digby and the Incomparable Mr Robert Boyl' (I, p. 377).[19] Traherne was not merely a poet and a devotional writer occasionally commenting on areas outside of his expertise. He responded to the contemporary vogue of atomism by developing his own, sophisticated atomic philosophy, and used his reflections on the indivisible particle to work out the paradoxes in his theology.

---

[17] Finn Fordham, 'Motions of Writing in the Commentaries of Heaven: The "Volatilitie" of "Atoms" and "ÆTYMS"', in *Re-reading Thomas Traherne: A Collection of New Essays*, ed. Jacob Blevins (Tempe, 2007), pp. 115–34 (p. 116).

[18] James J. Balakier, *Thomas Traherne and the Felicities of the Mind* (Amherst, 2010); Chad Michael Rimmer, *Greening the Children of God: Thomas Traherne and Nature's Role in the Ecological Formation of Children* (Eugene, 2019).

[19] As previously mentioned, Traherne took thorough notes on Henry More's *Divine Dialogues* (London, 1668) in his commonplace book. For more information on Traherne's reading of More, see Marks, 'Thomas Traherne and Cambridge Platonism'. Traherne dedicates the majority of a notebook to thoughts on Ficino, now known as the 'Ficino Notebook' (British Library, Burney MS 126). See Carol Marks Sicherman, 'Traherne's Ficino Notebook', *Papers of the Bibliographical Society of America*, 63 (1969), 73–81. Notes from Francis Bacon's *De augmentis scientiarum* appear in the 'Early Notebook', Oxford, Bodleian Library, MS Lat. Misc. fol. 45.

His notes on atomic philosophy owed much to Epicureanism, though with some alterations and reservations. Traherne cites lengthily from Theophilus Gale's *The Court of the Gentiles* in his commonplace book, where he paraphrases Gale's overview of the pre-Socratic history of atomism in his section on the 'Atom'.[20] Gale's primer of classical philosophy influences Traherne's own encyclopaedic works, with references to *The Court of the Gentiles* scattered across the *Commentaries* and *The Kingdom of God*.[21] His chapter 'Of Epicurisme' introduced Traherne to the theory that Epicurean atomism had a Judaic origin, through the mysterious figure of Mochus the Phoenician: 'the first great assertor of *Atomes* was *Mochus*, that famous Phenician Physiologist, who traduced them from the Jews'.[22] Mochus came to be associated with the Israelite Moses by scholars as prominent as More and Robert Boyle.[23] If Gale et al. convinced Traherne of a conceivable Judeo-Christian influence in the history of atomism, he would have discovered further sympathies between Christianity and Epicureanism upon reading Book X of Laertius's *Lives and Opinions*. This was one of the key sources for introducing Epicurean principles to early modern Europe. Laertius summarises the Epicurean 'Wise Man' as 'happy, though he should be tormented; [he is] the only person who will

---

[20] It is clear that Traherne conducted research into Epicureanism – his commonplace book contains notes on the history of atomism via Leucippus, Democritus and Epicurus, copied from *The Court of the Gentiles*. See 'Atom', fol. 18^b.

[21] For example, in the *Commentaries* Traherne begins his section on 'Amendment' by acknowledging the 'learned Gale an English Minister, to whom we are much obliged for several Pieces of Antiquity in this Discourse'. *Works*, III, p. 47. Traherne also refers to *The Court of the Gentiles* in his commentary on Aristotle (III, p. 188).

[22] *The Court of the Gentiles*, 'Of Epicurisme', pp. 440–48 (p. 443). Traherne copies this passage directly in his commonplace book, under the heading 'Epicurisme', fol. 42r.

[23] Robert Boyle notes that the atomic philosophy 'is by Learned Men attributed to one *Moschus* a *Phoenician*' in *The sceptical chymist; or, Chymico-physical doubts & paradoxes touching the spagyrist's principles commonly call'd hypostatical* (London, 1661), p. 120. Henry More makes the connection with Mochus in *Conjectura Cabbalistica* (London, 1653), p. 82. See also Introduction, p. 28.

return Thanks to his Friends, both present and absent'. This strikes a chord with Traherne's doctrine of 'Felicitie', the state of spiritual perfection and knowledge at which he directed all his writings.[24] From here, it is no small leap to find a connection between Traherne's Christian ethics and the Epicurean felicitous wisdom. Gale even refers to 'Felicitie' when defining the Epicurean Good of 'Pleasure' in *The Court of the Gentiles* as 'Indolence of bodie, & Tranquilitie of mind'.[25]

While 'Tranquilitie of mind' may have held some benefit for Traherne, he would not have approved of other Epicurean theories. He admits in the *Commentaries* that the Epicureans had been mostly 'Excellent, and Successfull in their Doctrine', but is swift to criticise and adjust problematic principles in order to bring the philosophy into closer harmony with his faith. In particular, he identifies a fundamental 'Disgrace': 'That any thing should hav a Being which it owes not to the Deitie, it is monstrous to conceiv' (III, p. 351). Traherne followed Gassendi and Walter Charleton in placing atoms under the command of a supreme deity who, he explains in *The Kingdom of God*, 'by those things that are next to Nothing ... so wonderfully composeth this Beautifull World, Effecteth Innumerable Works, and Conducteth them by his Wisdom to the Highest Attainments' (I, p. 349). Those 'things that are next to Nothing' are, of course, atoms. This reminder of the particle's slight physical status recalls the hierarchy established in Traherne's second 'Atom' poem, where the immaterial 'Nothings' of souls are supported by the indivisibles 'next' to them in the scale of metaphysical activity.

---

[24] Diogenes Laertius, *The lives, opinions, and remarkable sayings of the most famous ancient philosophers written in Greek, by Diogenes Laertius; to which are added, The lives of several other philosophers, written by Eunapius of Sardis; made English by several hands* (London, 1696), Book X, p. 258. This was the first English translation of Laertius's *Lives* to include his book on Epicurus. Traherne directs all of his writings at the obtainment of 'Felicitie', a state of spiritual perfection and omniscient knowledge. He defines 'Felicitie' in the Dobell manuscript poem 'The Vision' as the wondrous ability 'From One, to One, in one to see *All Things*'. See Ross, *Works*, VI, p. 15.

[25] Gale, p. 444.

His partial, Christianised acceptance of Epicurean atomism is so far conventional, but other comments reveal something more unusual. The closing 'Observations' to the atom commentary argue that 'The Epicureans' were 'absurd in Making Atoms distinct in Size and Figure' (III, p. 351). For Traherne, contra Gassendi and Charleton, an atom cannot have geometrical shape as such a figure must, by necessity, be divisible. He remains unspecific on the exact form of his atom, aligning it with a 'Mathematical Point' in both the *Commentaries* and *The Kingdom of God* (III, p. 339; I, p. 350). In the latter, however, he contradicts himself in a note on the ability of atoms to penetrate, and reside in, the smallest pores:

> For Atoms having no Dimensions, cannot by their Concurrence, tho Implicit Roundness is their Conceivable Figure, leav Spaces between them, when they touch each other. (I, p. 346)

Quite what Traherne means by 'Implicit Roundness' is hard to fathom. The clue appears to be in the allusion to the shape as their '*Conceivable* Figure'. In order to imagine an atom, it is necessary to conceive of a shape, and Traherne argues that 'Roundness' is the most appropriate form. This unusual conception of atoms as identical and round places him in good philosophical company, forging parallels with the metaphysical atomism of Giordano Bruno from the previous century and the theories of Gottfried Wilhelm Leibniz in the decades to come. While it is fair to conclude that Epicureanism introduced Traherne to the atomic concept, his developing theories owed just as much to his reading on and of Aristotle, Neoplatonism and the hermetic corpus, in addition to the Platonised atomism of Henry More.[26] Given the close association Traherne forges between indivisible atoms and souls, 'Implicit Roundness' was likely to please as a figure in which to conceive the particle. The figure of roundness aligns the atom with God,

---

[26] He was likely to have encountered the suggestion of roundness in Aristotle's *De Anima*, which opens by introducing the arguments of the author's predecessors. Aristotle acknowledges that Democritus's '"forms" or atoms are infinite in number', but describes only one particular form amongst many – 'those which are spherical he calls fire and soul'. Aristotle, *De Anima (On the Soul)*, trans. Hugh Lawson-Tancred (London, 1986), 403b31–404a5.

bringing it close to Cusanus's notion of the divine universe as 'a machina mundi whose centre, so to speak, is everywhere, whose circumference is nowhere, for God is its circumference and centre and He is everywhere and nowhere'.[27]

The subject of this chapter is the divinity of Traherne's atoms and their necessary role on his quest to become 'Expert in the Chiefest Mysteries of GOD and Nature' (III, p. 333). Specifically, my focus is on the emergence of an 'atomic poetics' in Traherne's works, which manifests as though in direct response to the challenges of recognising the 'Conceivable Figure' of indivisible substances. Traherne's poetic forms realise the spiritual productivity of the atom in a way not captured by his lengthy, exhaustive prose structures. As the second 'Atom' poem quoted at the top of this chapter demonstrates, it was his poetry that could preserve the *active* relationship between atom and soul in its recreation of a metaphysical mystery. Traherne's emphasis on the importance of 'Act' throughout his theology meets its appropriate medium in the form of the lyric, which was able to access 'Felicitie' and higher spiritual meaning through the fluidity of its movement.[28] Contrarily to the linear direction of his prose, his poems capture and embrace the differences in paradox, transforming their tensions into effective messages; his lyrics are able to flit between different subjects and outlooks, adopting the Cusanian perspective of 'everywhere' and 'nowhere' in their tendency to zoom in and out between microcosmic and macrocosmic states of experience. There are strong sympathies between these aspects of Traherne's poetic style and the features of his atomic philosophy. The seemingly omniscient, connecting power of 'Felicitie' that Traherne's poems enact – the ability 'From One, to One, in one to see All Things' ('The Vision') – also describes the motion of his atoms, as they seek to 'serv' creation and souls (III, p. 363).

Over the following sections I examine these connections and the productivity of Traherne's poetic spaces in exploring the spiritual necessity of the atom. I begin by expanding further on his philosophy of the 'Atom', an entity which, like the intended final version of the

---

[27] Nicholas Cusanus, *Of Learned Ignorance*, trans. Fr. Germain Heron (London, 1954), Book II, Chapter 12, p. 111.
[28] Traherne's commentary on 'Act' can be found in *Works*, II, pp. 170–87.

Commentaries, encompasses 'ALL THINGS'. From there, the second section considers the bond between atom and soul and the medium Traherne adopts to strengthen the relationship: poetry, a part-spiritual, part-physical form with the active power to express higher truths. Through the accommodating purposes of the atom, the capacious soul and the comprehensive poem, Traherne strives to make 'Great, Endless, External Delights' available to mortal embodiment.[29]

ATOMS AND ACT

Traherne followed More in associating knowledge of the atom with understanding of God. He agreed with More's theory of utmost indivisibility, holding the opinion in common with the Cambridge Platonist that even 'Almighty power cannot divide what is Absolutely Indivisible' (*Kingdom of God*, I, p. 432). Traherne also shared an inclination towards vitalism with the Cambridge thinker, though his had less to do with 'congruity' and was more a reinterpretation of Neoplatonic emanation, in which he was influenced by William Harvey's theory of blood circulation.[30] His atomic musings responded to More's metaphysics but they were not imitative. Both thinkers held that the atom nurtures the divine but developed different theories of accommodation. Within More's Neoplatonic cosmos, the indivisible atom takes form as an object of reflection, an analogy for the indivisibility of God and the inner life. Traherne's atom holds more prestigious and cherished a place in the Christian universe. Because atoms are indivisibles, he argues that their form is at one with their essence, or subject: they have 'an intire Essence evry one in its Distinct Existence' (*The Kingdom of God*, I, p. 342). This is a bold claim. As Traherne points out only a few paragraphs earlier, following Aristotelian metaphysics, 'even in things Spiritual there is a Distinction between the Essence, and the Existence of Evry object; the Being, and the Manner of its Being; the Subject it

---

[29] *Centuries*, I, First Century, aphorism 19.
[30] In *The Kingdom of God* Traherne suggests that as 'by the circulation of the blood (lately found out) all Life and Motion is maintained: This in the Microcosm is answered with an Universal Circulation in the Macrocosm: the Sun being as it were the Heart of the Univers' (I, p. 350).

self, and its Qualities, or Accidents' (I, p. 341). If atoms, unlike other physical and spiritual entities, concentrate their subject and form into one unified mode of being, they represent the closest things in creation to God – the deity evoked by More as the Pythagorean 'Unmoved Monad'.[31]

More proposes the accommodative atom as the groundwork of a Christian reading of the natural world. Traherne views it as the end of knowledge and spiritual experience itself. His emphasis is on building empathy with the divine capacity of the atomic interior; something that can only be felt and known. He describes how the atom forms a vessel for divine information at an earlier stage of the commentary:

> yet is there an unsearchable Abyss of Wonders contained in it; innumerable Difficulties, Uses, Excellencies and Pleasures concentering in its Womb, for our Information and Happiness; the clear Knowledge of which will make us Expert in the Chiefest Mysteries of GOD and Nature. (III, p. 333)

The 'Wonders' contained within the atom are 'innumerable' and piled within 'an unsearchable Abyss'; yet they exist for 'our Information and Happiness' and 'clear Knowledge' of them is considered possible. Framed as the ultimate object of enquiry, Traherne's 'Atom' takes on an extraordinary level of importance as it provides a way to develop expertise in 'the Chiefest Mysteries of GOD'. What are the 'Wonders' encased within an atom, and how can they be 'unsearchable' and yet clearly known all at once?

The answer is in the inextricable link between the atom and the comparably wondrous act of being human. Traherne suggests in his *Commentaries* that the atom is 'A good Emblem of Humility' (III, p. 345), redirecting More's atomic 'embleme of the Deity' to focus on human experience.[32] His reasons for viewing it thus are distinct from those of his contemporaries who discuss particles with distaste for their relation to dust and worthlessness.[33] The

---

[31] *Psychathanasia*, 3.3.12, in Henry More, *Philosophicall poems* (Cambridge, 1647).

[32] More, 'The Interpretation Generall' in *Philosophicall poems*, p. 430.

[33] Atoms commonly featured in written insults from the period, specifically those that assert the worthlessness of the mortal human body by the frailty of its material construction. For example, Zachary Coke remarks on how

description of the emblem as 'good' is noteworthy. Traherne is enamoured with the idea of an atom. He observes its material poverty not with despair but with joy in response to its 'great Miracle in a litle Room' (III, p. 333).[34] Atoms contain an 'Abyss of Wonders' despite and indeed because of their meek physical form. More than a symbol for the Christian recognition of human frailty, and quite divorced from threat, Traherne's atoms demonstrate the good that a life of humility can achieve. The very reason the atom has become a symbol of humility, Traherne explains, is because 'the least things are the most secure' (III, p. 345). His indivisible particle is a model for the ideal human individual, with emphasis not on the Lucretian 'swarm' but on the stable immutability of a single atom amidst surrounding change.[35] Its essence remains the same regardless of how it moves in the flux of the wider world. As Traherne writes in the second poem at the end of the commentary, despite the 'varietie' in sublunary things, the estates of atoms 'Are still the same: And which is very Strange, / Without Change in themselvs, they all things change' (III, p. 358). The blessed human being is as constant in faith as the unchanging interiority of an atom and his or her spiritual capacity is comparably mysterious.

---

man is 'but a graine or Atome'; Joseph Hall shows further disgust in his *Select Thoughts* when he considers 'what an insensible Atome man is, in comparison of the whole body of the Earth'; in John Corye's comedy *The Generous Enemies* (1672) a character receives the biting remark, 'You heap of Atoms kneaded into flesh!' Zachary Coke, 'Address to "THE ILLUSTRIOUS, His Excellency Oliver Cromwel"', in *The art of logick; or, The entire body of logick in English* (London, 1657), sig. A3; Joseph Hall, *Select Thoughts, or Choice helps for a pious spirit* (London, 1654), p. 74; John Corye, *The Generous Enemies or The Ridiculous Lovers* (London, 1672), p. 61.

[34] Traherne expresses similar wonder for other small objects in his study of 'ALL THINGS' – notably the 'Ant', another entity he describes as 'a great Miracle in a litle room' (III, p. 93).

[35] For a Lucretian account of the tumultuous movements of atoms in the state of 'chaos', see *The Works of Lucy Hutchinson, Vol. 1: The Translation of Lucretius*, ed. Reid Barbour, David Norbrook and Maria Zerbino (Oxford, 2012), Book 5, lines 448–66 (p. 325).

Here and throughout Traherne's works, accounts of atomic motion connect with the significance of act and activity in his theology. He describes 'Moving Atoms' as 'the Spirits of the World that Actuat all things, or Act in all things' (III, p. 349). Elsewhere in the *Commentaries of Heaven*, 'Act' is granted its own section, another long entry that is followed closely by separate commentaries on 'Action' and 'Activity'. The nature of 'Act' is so important, Traherne explains, because God is in essence 'a Voluntary and Eternal Act, most Pure and Simple' and, 'having no substance but what the Act is, begot Himself by Acting' (II, p. 184). While the deity is a 'Simple Absolut Act' (II, p. 171), the human individual shares in this power by containing the potential for divine act, which must be set in motion to be realised. Without active contemplation, Traherne muses, we cannot register what is innately within us: 'How all Eternity is set before us we can tell, but how it is seated in us we cannot … Perception being the Greatest Riddle in the whole world, the Greatest Abyss, the Greatest Mystery' (II, p. 183). By 'Perception' in this context, Traherne refers to the act of apprehending not the observed 'Eternity' of external creation, but the greater mystery of the equally infinite, eternal rational soul within – the very substance of the contemplating mind itself. This 'Riddle' is solved in Traherne's theology by activating spiritual enquiry. As James J. Balakier has written on Traherne's 'Notion of Act' in contemplative practice, referencing the *Select Meditations*, '[h]is attitude is always that, no matter how incredible something may sound, "Experience will make it Plain".'[36] Nowhere is this more the case than with the nature of the soul, which takes form to the enquiring mind through the process of discovering and experiencing its own capabilities. Benjamin J. Barber recognised this reflective act of self-knowledge with his claim that, '[f]or Traherne, the act of contemplation is both the soul's occupation and its essence.'[37] As long as the soul is actively

---

[36] James J. Balakier, 'Felicitous Perception as the "Organizing Form" in Thomas Traherne's Dobell Poems and Centuries', *XVII–XVIII: Revue de la Société d'études anglo-américaines des XVIIe et XVIIIe siècles*, 26 (1988), 53–68 (57).

[37] Benjamin J. Barber, 'Syncretism and Idiosyncrasy: The Notion of Act in Thomas Traherne's Contemplative Practice', *Literature and Theology*, 28.1 (2014), 16–28 (25).

engaged with itself and exercising its potential, the great 'Abyss' of spiritual perception is revealed and made tangible. Traherne continues in the 'Act' commentary: 'when our Souls are Exerted, all our Powers being turned into Act, we becom like Him infinit Spheres, He in us, and we in Him, all Sight, all Esteem, all Love, all Joy etc. being Pure Act as He is' (II, p. 183).

Both the Cusanian reference to 'infinit Spheres' and, especially, the allusion to perception as the 'Greatest Abyss' connect with Traherne's understanding of the atom, the 'unsearchable Abyss of Wonders'. Traherne's atom, like his theory of the soul, is filled with divine perception that is simultaneously – and paradoxically – hidden from yet not only available but essential to human enquiry. His 'Abyss' may refer primarily to a bottomless, unfathomable pit or chasm, but the word carries the additional sense of the 'primal, formless chaos' out of which all things were created.[38] This definition brings the act of seeking knowledge and spiritual understanding directly into contact with the atomic particle, which took its form from the original 'formless chaos'. Contemplation of the atom, this suggests, could inspire direct revelation of a great 'Mystery' in perception: of how created, distinct entities emerge from unformed, inactive potential. It is only through active spiritual reflection that understanding becomes possible.

Traherne reflects on this in his commentary on the 'Atom'. Its 'Abyss' is 'unsearchable' until it is experienced, which releases its significance from potential to active being. Divine power fills the atom with eternity, because God 'lodgeth his whole DEITIE, / In one small Centre, yet is not confined' (*Commentaries*, III, p. 339). Atoms are endowed, therefore, with the eternal and omniscient perception of God. The challenge is in how the human individual can access this knowledge. Prior to this conclusion, Traherne informs his reader that the atom is the very key to be applied in unlocking its own secrets, describing the particle as a 'means wherof all Things Past, Present, and to Come, may at once appear before the Understanding' (III, p. 338). The word 'means', common within the terminology of reformed religion and hearkening back to the recommended use of 'ordinary meanes' in the

---

[38] *OED*, 'Abyss, n. 1.a'.

Westminster Confession of Faith, suggests that the believer should use the atom as a personal object of assurance.[39] There is of course nothing 'ordinary' about the means presented by the indivisible particle, which, according to Traherne, provides a point of contact between the 'Understanding' and omniscience. The atom, itself requiring a 'Conceivable Figure' to become fathomable (I, p. 346), becomes an important object for the soul to embrace, enter and absorb on its quest to realise its own form, and the substance of divine 'Act', through active contemplation.

As Traherne admits, it is only possible to recognise the 'Obscure ... Simplicitie' of the atom through 'Intuition':

> I know not how, by reason of its Simplicitie it is Obscure, and inexpressible, and yet for that reason is the most easy to be apprehended in the whole World. Multiplicity breeds Distraction, but Singleness and Indivisibilite contract our Thoughts, and the Things are if not Known by Causes, felt by Intuition. And what we feel in an Atom (with our Minds) is the Matter of it. (III, p. 341)

He emphasises the atom as a point of contraction for the enquiring mind, ready to reveal divine truths through its 'Simplicitie'. Simplicity is one of the most used nouns in Traherne's writings. Chad Michael Rimmer has recently offered a definition: 'The atom provides the material basis for a common nature shared by all creatures. "Simplicitie" is the term for that common nature.'[40] Rimmer correctly stresses the importance of empathy and a common nature across Traherne's writings, but there is more to the significance of 'Simplicitie'. Here and elsewhere, the word refers to a

---

[39] See especially the eighteenth chapter of the *Westminster Confession of Faith*, which defends the possibility that the 'true believer' may 'without extraordinary revelation, in the right use of ordinary means', attain 'infalliable assurance'. The most common means of grace and assurance, the *Westminster Confession* outlines, are meditation on the Word, participation in the sacrament and devotion to prayer. *The humble advice of the Assembly of Divines, now sitting at Westminster, concerning a Confession of Faith* (London, 1647), pp. 31–32, 1–6, 37–38, and p. 47.

[40] Rimmer, p. 74.

purity or simplicity of form.[41] A clear distinguishing factor between Traherne's interpretation of the atom and other atomic theories of the century, including More's, is his insistence that atoms are material but not corporeal. To recall the second 'Atom' poem from the *Commentaries*, these indivisible particles are 'simple ones'. As he writes earlier in the section: 'An Atom being infinitly small is Incorporeal. For since the Essence of a Body is to be Extended, and to have Parts out of Parts, an Atom can be no Body' (III, p. 350).[42] It would be a contradiction for the indivisible atom to have extension, so Traherne reasons it cannot have body (as bodies come with parts). He defines it in *The Kingdom of God* as 'a Seed of Corporeitie', of 'No Dimensions' itself but active in its role as an instrument of vitality (I, p. 342). It is 'Moving Atoms', we recall, that 'Actuat all things'. These particles, therefore, are of the utmost 'Simplicitie' in form, but become 'Obscure' and 'inexpressible' in attempts to describe them.

The mind – simultaneous with soul in Traherne's writings – needs to contract itself to the point of 'Simplicitie' in order to apprehend the atom's mysteries. Its method for doing so is to feel 'by Intuition'. Intuition is knowledge or belief that is immediate, instinctive and unaffected by rationality or physical perception. In other seventeenth-century texts, references to 'intuition' are often paired with the words 'bare', 'naked' and 'open' (amongst many others by John Locke and Richard Baxter, respectively) and understood to offer the clearest, fullest perspective of holy and distant objects.[43] The common theme, however, is that intuition – the method of attaining

---

[41] See also, for example, the reference to 'Simplicity' in the Burney MS poem 'Eden', which compares the state of childhood innocence with Adam's prelapsarian purity: 'My happy Fate / Was more acquainted with the old / And innocent Delights which he did see / In his Original Simplicity.' *Works*, VI, p. 93.

[42] His stance here is notably different to More's take on the atom, or 'physical monad', to which he assigns extension. More preferred the term 'physical monad' from 1662. See Jasper Reid, *The Metaphysics of Henry More* (Dordrecht, 2012), p. 51.

[43] Examples of these recurring images can be found in John Locke, *An essay concerning humane understanding* (London, 1690), p. 264; John Smith, *Select Discourses* (London, 1660), p. 97 and Richard Baxter, *A saint or a brute. The certain necessity and excellency of holiness* (London, 1662), p. 319.

knowledge belonging to God himself – is too divine an activity for mortals restricted by the imperfect faculties of their fallen bodies.[44] Unsurprisingly, Traherne was not of this opinion. The process of feeling 'by Intuition' enables his curious soul to imitate, recreate and join the divine 'Act' of creation. Our minds not only 'feel' outwards but, crucially, turn inwards: we can 'feel in an Atom (with our Minds)' because we contain the very objects we seek. Recalling Rimmer's definition of 'Simplicitie' as 'common nature', all human beings are formed from atoms, so the particle is something we come to know in the same manner we come to know ourselves, our souls, and God. Indivisibility, Traherne argues, is innate to our very existence – the material construction of our bodies, the immortality of our souls and the omnipresence of the divine. The advice to 'feel inside an Atom', therefore, is beyond corporeal experience but more direct than mere metaphorical analogy. As the epitome of 'Singleness and Indivisibilitie', the atom serves as a point of contraction to lead the devout enquirer to the first cause, the deity that 'lodgeth' there.[45]

This raises a further question as to what Traherne intends by 'the matter' of the atom, if its contents are the spiritual essence of divine power. While his particles are incorporeal, they are nevertheless still a part of – and the founding basis for – material creation. Later in the commentary, he classifies the atom as a 'Material Spirit', a title that he also attributes to rays of light and 'Spirits' that 'move

---

[44] The jurist Matthew Hale summarised this opinion: 'our knowledge is not by Intuition, as the Divine Knowledge is'. Matthew Hale, *A discourse of the knowledge of God, and our selves* (London, 1688), p. 45. For another example, see John Hall: 'we are not able to know or apprehend truly our selves, as the divine essence is, by means of continual presence and intuition'. *Of government and obedience as they stand directed and determined by scripture and reason* (London, 1654), p. 261.

[45] This recalls the parable of the mustard seed, from Matthew 13. 31–32: 'He told them another parable: "The kingdom of heaven is like a mustard seed, which a man took and planted in his field. Though it is the smallest of all seeds, yet when it grows, it is the largest of garden plants and becomes a tree, so that the birds come and perch in its branches."' I am grateful to Sarah Brown for making this connection, 03 October 2019.

in the Blood' (III, p. 351).[46] He then explains the location of these 'Material Spirits' within his universe:

> Material Spirits are the Mean or Clasp, between Immaterial Spirits and Material Bodies. for with Spirits they are void of Body, with Bodies they are full of Matter. (III, p. 351)

Traherne hints at the information atoms have in store and how they present as 'means' for the faithful Christian. God created rays of light to reveal his presence and establish a point of contact between the terrestrial and the divine; 'spirits' within the blood enable the mingling of body and soul within the human individual. The atom also is a 'Mean or Clasp' between the spiritual and the corporeal. This is related but different to More's 'vehicles', the matter of 'vital congruity'. Atoms, unlike the vehicular bodies of More's cosmos, are not physical vessels for spirit but entities perfectly poised in between material and spiritual states. The intermediary presence of the 'Material Spirit' unites body and soul into one indivisible realm of experience. With similar qualities to Ficino's cosmic 'body that is not a body', Traherne's atoms preserve the harmonised spiritual and material activity of the natural world.[47]

They are 'full of matter', therefore, not just from their status as the 'seeds' of physical creation, but because of the import of their spiritual information. Through its intermediary status, the atom is responsible for establishing the relationship between spiritual life and matter that makes knowledge of the divine 'Act' possible. The atom unveils the mysteries of divine perception, its

---

[46] For more on the philosophy of medical spirits and the relation to materialism, see D. P. Walker, 'Medical Spirits in Philosophy and Theology from Ficino to Newton', in *Arts du Spectacle et Histoire des idées* (Tours, 1984), pp. 284–300.

[47] Both More and Traherne were influenced by Marsilio Ficino's concept of an intermediary matter between 'gross' body and the divine soul. Ficino writes that 'the aid of a more excellent body – a body not a body, as it were – is needed. We know that just as all living things, plants as well as animals, live and generate through a spirit like this.' See *The Book 'On Obtaining Life from the Heavens'*, in *Three Books on Life: Marsilio Ficino*, ed. and trans. Carol V. Kaske and John R. Clark (Binghamton, 1989), pp. 236–405 (p. 257).

'unsearchable Abyss', through acts of connection – and, crucially, the active enquiries of another perceptive agent become necessary to gather the information. As Rimmer has noted, 'it is clear that Traherne understood something rather profound about the nature of an atom. He maintained that the properties of an atom require a relationship to display their qualities.'[48] By 'relationship', Rimmer refers to the strength 'of a conglomeration or community of atoms', through which the individual particle, unfathomable on its own terms, joins with others to take form.[49] The strength of atoms in multitude was undeniably important to Traherne, but there is another, more significant relationship through which atoms 'display their qualities' and grant divine knowledge. It is through the connection between the enquiring, active subject – Soul – and the capacious object, filled with spiritual information – Atom – that the object comes into being, taking form and revealing its mysteries through active contemplation. Across his writings, it was through the medium of poetry that Traherne most productively supported and explored this relationship.

ATOMS, SOULS, POEMS

The focus of this section is on tracing a poetics of the atom within Traherne's writings. As I have just explored, it is through active contemplation that his atom takes its form as a vessel for spiritual knowledge. Through lyric poetry, Traherne set this divine 'Act' of understanding in motion: his poetry is characteristically *active*, moving between and uniting past, present and future experiences. This is exemplified by the running theme of the Dobell and Burney manuscript lyrics, which reflect on the poet's experiences of childhood bliss, not as distant memories of an innocence now lost, but as ideal states of being present with and accessible to the adult soul.[50] Traherne's poems, like his atoms, present a 'means wherof all

---

[48] Rimmer, p. 74.
[49] Ibid., p. 75.
[50] For an excellent study of Traherne's interpretation of childhood innocence, see Elizabeth S. Dodd, *Boundless Innocence in Traherne's Poetic Theology* (Farnham, 2015).

Things Past, Present, and to Come, may at once appear before the Understanding' (III, p. 338). An advantage of poetry over prose was that, while the lengthy essays of the *Commentaries* could examine the details of complex theories, Traherne's lyrics could encapsulate and perform them, thereby putting into practice the very 'Act' sought in description. These poetic spaces provide a dynamic focus for contemplation and, like atoms, actively 'serv' souls and the divine impulse behind their creation.

The technical elements of Traherne's poetry – rhyme, punctuation, the repetitive sounds that characterise his most passionate moments, and the prominent biblical resonance of his metaphors – come to accommodate the comprehensive 'Act' of the soul. In the notes under the heading 'Poett' in his commonplace book, Traherne affirms the suitability of poetic art for expressions of divine experience, writing that the poet, more so than any other human being, 'hath his will all wayes turned to such an high Key, or straine, as ordinary wits cannot reach'.[51] Traherne copies his information for this section from the Platonist Thomas Jackson's 1625 *A treatise containing the originall of vnbeliefe*, a study wherein the writer aims at 'rectifying our belief or knowledge' in God.[52] The chapter from which the following quotations are taken instructs the reader as to how love of the deity leads to understanding through 'internal illumination', and within his explanation Jackson offers an analogy of the 'most delicate' poetical inventions originating from the poets 'most in loue' with their subjects.[53] For Traherne, poetic skill is a direct means to the communication with and of God, as his selective note-taking from Jackson implies. As for George Herbert, poetry is an expression of internal reflection and devotional

---

[51] See the section marked 'Poett' in Traherne's commonplace book, fol. 76.

[52] Thomas Jackson, *A treatise containing the originall of vnbeliefe* (London, 1625). In Jackson, the text copied into the commonplace book is found on pages 185 and 187 (Chapter XX, 'Of the speciall nutriment which the Poetrie of auncient times did afford to the forementioned seedes of Idolatrie') and pages 461–62 (Chapter LI, 'The best meanes to rectifie and perfect our knowledge of God is to loue him sincerely').

[53] Ibid., pp. 460 and 461.

transcendence.[54] It is the work not of worldly 'nimbleness of conceit or apprehension' the commonplace book explains, quoting Jackson, 'but *the* unrelenting temper of imbred desire, & uncessant sway or working of secret Instinct, w*hi*ch brings *the* seeds of Knowledge to just truth growth & maturitie'.[55] Jackson's vocabulary depicts poetic skill as constant and apparently routine, not primarily the result of an exterior inspiration or 'apprehension' but the reproduction of the 'imbred' rhythms and 'desire' of the heart, with the ongoing end of growing through 'Knowledge' to spiritual 'maturitie'. The link between poetry and intuitive, 'secret Instinct' may have been what drew Traherne to include so many verse devotions in his philosophical studies, where, as his conclusion of the instrumental value of 'Intuition' in the atom commentary suggests, the act of *feeling* with the understanding invites an intimacy between the soul and the object it seeks. It is in the poetry that accompanies his philosophical prose that he plants '*the* seeds of Knowledge' – in Jackson's words – to reap understanding like a plant of '*the* soundest rootes, & sappiest stemmes', continually engaged in spiritual growth.[56]

Though Traherne wrote dozens of pages of prose for every poem in his oeuvre, his poetry held special purpose as a physical medium equipped to observe divine entities.[57] Opening the series of 'Poems of Felicitie' in the Burney manuscript, Traherne describes in a versified address 'From the Author to the Critical Peruser' how poetry ought to fulfil its function of accommodating spiritual truths in material form:

> The naked Truth in many faces shewn,
> Whose inward Beauties very few hav known,
> A Simple Light, transparent Words, a Strain
> That lowly creeps, yet maketh Mountains plain,

---

[54] Traherne copies Herbert's poem 'To all Angels and Saints' into his 'Church's Year Book', Oxford, Bodleian MS. Eng. th. e. 51, fol. 112.
[55] Traherne, commonplace book, fol. 75v.
[56] Ibid.
[57] Traherne's brother Philip also understood this spiritual purpose, prefacing the manuscript of 'Poems of Felicitie' with a dedicatory poem in which he praises the 'Sacred Relicks' Thomas left behind and poetry more generally, writing 'DIVINITY / And POETRY / We call Our Gifts: Indeed they are Thine Own: / Theses Faculties from Thee do flow'. *Works*, VI, p. 83.

> Brings down the highest Mysteries to sense
> And keeps them there; that is Our Excellence:
> At that we aim; to th' end thy Soul might see
> With open Eys thy Great Felicity,
> Its Objects view, and trace the glorious Way
> Wherby thou may'st thy Highest Bliss enjoy.[58]

The proposed transparency of his verse owes much to the 'plain' style of George Herbert, who is indirectly but prominently referenced in a forthcoming rejection of 'curling Metaphors that gild the sense'.[59] Within tight, regular heroic couplets, Traherne assigns poetry the purpose of communicating pure truths, corresponding the mysteries within – 'inward Beauties' – with those without, the 'highest Mysteries' supposedly inaccessible to sensual perception. Fittingly for a collection entitled 'Poems of Felicity', Traherne describes the act of 'Felicitie' itself, as executed through poetic communication: the act of seeing one's soul and 'its Objects', the divine information housed inside it. Traherne's advice that the reader may 'trace the glorious Way' to 'Highest Bliss' in his poems recalls a claim in the *Centuries* that the 'Image of God' can be known by the 'lineaments' of the soul.[60] Reading the lines of his verse recreates the act of 'tracing' the divine lineaments that Calvin had argued were 'so vitiated and maimed, that they may truly be said to be destroyed', but which Traherne restored through his doctrine of insatiability.[61]

'From the Author to the Critical Peruser' does not mention atoms directly, but his language is reminiscent of the indivisible particle which, as a means for understanding the divine 'Act' of creation, also

---

[58] *Works*, VI, p. 84. The manuscript is in the British Library, Burney MS 392.
[59] The phrase 'curling metaphors' was likely to have been lifted from 'Curling with metaphors', 'Jordan II', line 5. Gladys Wade was the first to make this observation. See *The Poetical Works of Thomas Traherne*, ed. Gladys Wade (London, 1932), p. 268.
[60] *Centuries*, I, First Century, aphorism 19.
[61] See Jean Calvin, *Commentary on Genesis*, on the image of God in man: 'although some obscure lineaments of that image are found remaining in us; yet are they so vitiated and maimed, that they may truly be said to be destroyed'. *A Commentary on GENESIS: Jean Calvin*, ed. and trans. John King, 2 vols (Edinburgh, 1847), I, p. 95.

'Brings down the highest Mysteries to sense / And keeps them there.' There is a characteristic blurring in Traherne's language between what is sensorial and what is spiritual, or intellectual understanding. The physical features of his poetry are evident from the portrayal of truth in different 'faces', or from the 'strain / That lowly creeps'; a personification that emphasises a reliance on base matter even for the spiritual music of his poetic subjects. But the opening focus of the poem remains 'The Naked Truth', which Traherne proceeds to describe in its different 'faces' (as light; as words; as music) before finally announcing in the sixth line that it is 'Our Excellence'. As I discussed in the previous section, 'Naked' is an adjective that commonly accompanies references to intuition in seventeenth-century texts. The act of grasping internal truths in poetry recalls the notes from Jackson in the commonplace book and moreover Traherne's instructions in the *Commentaries*, to intuitively 'feel in an Atom (with our minds)' to apprehend its 'Simplicitie'. As a feeling, intuition continues to blur the qualities of material and spiritual enquiry, somewhere between a touch, an emotional response and an instantaneous recognition. Poetry enacts this process, Traherne tells us in 'From the Author to the Critical Peruser', to convert mysteries to human understanding. It 'maketh Mountains plain' from an earth-bound perspective, an act that implies both a lowering of the lofty peaks of enquiry and an ability to view 'highest Mysteries' in their unadorned 'Simplicitie'.

This 'Simplicitie' is a perspective of utmost clarity; an ability to grasp the importance of divine objects in their purest form and a recognition of the common nature between 'All Things'. As with the 'Simplicitie' of the atom, the substance and source of all meditation comes back to 'Act', which Traherne's poems pursue as a common theme.[62] In the *Commentaries of Heaven*, Traherne concludes his section on 'Act' with a poetic exploration of the topic, which appropriately aspires to define its subject by enacting it:

---

[62] The notion of 'Act' is explored especially in 'The Preparative', 'My Spirit', 'Fullnesse', 'The Anticipation' and 'The Recovery'. See *Works*, VI, pp. 11, 26, 30, 52 and 56.

> An Act! What is an Act? An Act Acted
> Is like a Sea that into the Ocean Shed.
> It is like the Sea, and yet tis not the same:
> And yet it is, but then hath lost its Name.
> The Sea that fell from off the Mountains yields
> The very Same that once vaild ore the fields.
> The very same but in another place.
> The very same but with another face. (II, p. 185)

The ever-flowing cycle of cause and effect suggested by Traherne's definition of 'Act' as the 'Deity' is put into action, literally set in motion in order to be experienced and understood at closer range. At first glance the poetic form, with its clear division of heroic couplets, appears regular. It is however not until the fourth line that the poem settles into iambic pentameter: the additional beats of the second and third lines, combined with the destabilising caesurae of the opening, create the impression of a form that is straining at its bounds. Through enjambment, repetition and metrical irregularity, the poem energetically resists its formal limits to demonstrate the active interconnectedness of 'All Things'. Each line leaps forward to consider act 'with another face', mysteriously 'The very same' yet different. In the movement from the first to second line – 'An Act Acted / Is like a Sea that into the Ocean Shed' – Traherne's vocabulary transitions from hard, immobile consonance to forward-moving sibilance. The visual and aural connotations of the simile perform the 'Act' in action. Anaphora in the closing lines stresses that there is a common essence to all things and their movements, embellishing the simile that 'An Act Acted' is like the cycle of water between sea, ocean, mountains and rain. Excessive repetition over the next few lines – 'is like', 'yet tis', 'but' and 'the very same' – builds the sense of importance in defining 'Act' through similitude, for it is in all things and is the creative source behind all matter. As a thing of glory, 'Act', like the sea moving constantly over the land and into oceans, takes on the form of the objects to which it belongs and which it empowers. It is the essence of the soul that imitates God and becomes 'Image and mirror of Him self' (II, p. 187).

Atoms are present, if unannounced, in these poems as entities – like the essence of 'Act' – that 'Without Change in themselvs, they all things change' (III, p. 358). The atom poems from the *Commentaries* demonstrate how contemplation of the indivisible particle could expand the reach of Traherne's active poetry, and how poetic form could enrich intuitive exploration of the atom in turn. This is illustrated explicitly in Traherne's attempts to observe the infinite journey of a single atom. The perspective he adopts for these enquiries parallels his pursuit of an 'Act' which, through its various appearances, remains '[t]he very same'. His volatile, continuously active – yet spiritually constant – atom is not only a part-material, part-spiritual enactment of the divine 'Act'; it behaves as though it were a soul undergoing the process of metempsychosis. Traherne's encyclopaedic writings establish a strong connection between atomic movement and metempsychosis, which stems from the supposed origin of atomism in Pythagorean philosophy (a theory he shared with Henry More). He observes these movements twice, in poetry and in prose, but it is poetic form that comes closest to realising the capabilities of the particle:

> It like an Angel all in Gold comes down
> And shooting through the Skin of man doth Crown
> It self with more Perfection; in his flesh
> And Blood it plays, and doth it self refresh.
> Or els perhaps it darts into a Gem
> And so is put into a Diadem,
> Or if it scapes from thence, and is not Worn
> By those whose Golden Crowns are lind with Thorn
> It mingles with the Baser Earth, and is
> A Spirit there, and having done in this
> Comes up into a litle Spire of Grass
> Which being eaten by som silly Ass
> Turns to a Part therof this silly Beast
> Is taken by a Lion… (*Commentaries*, III, pp. 361–62)

This one from the Earth may be carried into a Root, or Seed, and breath up at last into a Spire of Grass, be Eaten by a Beast, assist in the form of Nourishment, and pass into Flesh: that Flesh may be eaten by a Man, and become part of his, for a Considerable Season. (*The Kingdom of God*, I, p. 349)

Poetic form inspires the active voice in describing an atom, like a soul, in charge of its own destiny, insatiably seeking to become part of more objects. The movements of the poem are anything but linear: the atom descends '*like* an Angel' (italics mine; Traherne could be implying, once again, that atoms are more significant than the object of his simile) from where it explores the interior of the human body, before rising again to a crown; it then falls back to Earth, grows in grass and is consumed by an animal before, he later suggests, it could come 'By chance unto some Princes Table' (III, p. 362). In contrast to this dynamic rising and falling, the prose from *The Kingdom of God* traces a chain of absorption and digestion expressed in the passive voice (it 'may be carried'), with a distanced formality.

Traherne's poem closely follows the movements of an atom that assumes the properties of a soul, moving effortlessly across time and space. The artificial wealth of the 'Diadem' is contrasted with the greater riches of the human body, which, in spite of the atom's spectacular descent 'all in Gold' (another allusion to a precious stone, as well as references to fire, light and divine glory), crown 'It self with more Perfection'. As Traherne writes in the Dobell lyric 'The Person', the 'Sacred Riches' of 'Muscles, Fibres, Arteries and Bones / Are better far then Crowns and precious Stones' (VI, p. 40). His emphasis on the perfected status of the atom, with the mention of 'self' repeated between the third and fourth lines, reflects his belief in the divinity of material experiences. From there, the language of affective piety strengthens his empathy with the atom as a point of contact between matter and spirit. A 'Diadem' morphs into 'Golden Crowns ... lind with Thorn', alluding to the crown of thorns worn by Christ and a higher example of the merging of matter and spirit, the incarnation. Strikingly, Traherne's 'Crowns' are plural. One possible significance of this is that the infinite number of atoms within Christ's crown, now scattered, have entered 'All Things' – therefore enabling all beings to 'wear' an object, and experience a seismic event, transferred from the distant past into a continuous present. There is more than this, however, in the suggestion that each human being, a composite union of body and soul, empathises deeply with the incarnation and Passion of Christ. An ancient act is continuously re-enacted in the present through atomic motion and the movements of affective piety.

The single atom, like the soul, lays claim to 'All Things'. This intimacy recalls Traherne's musing on the combined human form from the Dobell Folio poem 'Wonder', which opens with the ecstatic declaration:

> How like an angel came I down!
> How bright are all things here!
> When first among his works I did appear
> O how their glory me did crown!
> The world resembled his eternity,
> In which my soul did walk;
> And ev'ry thing that I did see
> Did with me talk. (VI, p. 4)

The movements of the pure, infant soul in 'Wonder' are like the atomic motions of the *Commentaries* poem, to the extent that both atom and human descend 'like an angel' and both receive a crown. Traherne imagines his new-born soul crowned by the glory of God's 'works', with which his innocent capacity forms instant relationships. The atom, in turn, enters directly into an active relationship with the person, and in 'shooting through the Skin of man doth Crown / It self with more Perfection'. There is encouragement here to compare the two indivisible entities, atom and soul, both of which come 'first among his works'. What emerges is a clear interdependence between the two. Souls are crowned by their interaction with 'All Things', the common denominator of which is, of course, atoms. Likewise, atoms acquire greater 'Perfection' by contributing to the human body, the vehicle housing the soul. Contact between indivisible particle and rational soul is vital for the human being's desire to experience 'All in All' and to walk in the 'world [that] resembled his eternity'.[63]

Knowledge of Traherne's atomic theory informs a reading of his explorations of 'Act' in poetic form. The 'Naked Truth in many faces shewn' ('From the Author to the Critical Peruser') applies as readily to the ubiquitous atom as it does to the shifting utterances of the poem. Like the soul, the atom would become a 'Dungeon of

---

[63] 'All in All' is the title of a section in the *Commentaries*, III, p. 11.

Darkness' (*Christian Ethicks*) if it were not for its endless movements between earth, water, air and fire. The atom teaches by example that the inactive soul is cursed by 'Sin and Ignorance' (III, p. 354). It is necessary to keep acting, enacting and accumulating knowledge for universal good. So it is in the first of two 'atom poems' from the *Commentaries* that Traherne traces the progress of 'some small Atom from a Mirie Fen':

> As Earthly vapors by Celestial fire
> Refin'd, from Seas unto the Heavens aspire;
> Or some small Atom from a Mirie Fen,
> By shining Beams awakened, leavs its Den,
> And Mud behind; and rising on the Wings
> Of that Bright Beam, returning with it Springs
> Unto the Fountain, whence the River came
> And there absorpt becomes a fiery flame,
> Or Part of it...
> ... Even so from Slime,
> Ascending by the Spheres and Orbes of Time,
> I leav the Darker Prisons of the Womb;
> And to the Chambers of his Glory com:
> Transformd in Nature from polluted Mire,
> From duller Clay to quick Celestial fire:
> Bright and Volatile, Pure and Activ too,
> More Glorious made, by coming to the view
> Of all that see, then as Exalted there
> I Stood alone, tho in the Sun I were. (III, pp. 352–53)

The volatility of the poem and its subject parallels the journey of the atom that 'like an Angel' came down, but the emphasis here is on movement from the opposite direction. Like a soul in pilgrimage, the atom rises on the 'Bright Beam' of 'Celestial fire' to reach the heavens. Its joyous return to the 'Fountain' – into which it nimbly 'Springs', an active, onomatopoeic verb Traherne considered important enough to capitalise – could apply Neoplatonic imagery to symbolise a reunion with its divine source, the original 'Act' of the divine creator. His focus on act is emphasised further by the frequent use of 'or', which demarcates separate examples, or 'different faces', of the same act of ascension. The atom is one of a trio that provides a model for the ensouled individual, but it is the movements of the individual particle, rather than the aspiring seas or earthly vapours, that Traherne chooses to observe. Its innate volatility and inclination towards motion inspires the poet to draw a parallel with the glorious human being, who, like the atom, will be 'Bright and Volatile, Pure and Activ' in their state of perfection.

Traherne attests that it is the duty of the indivisible particle to follow this inclination, emphasising that the 'Noble' atom is one that continues moving and ministering to other beings and objects in the world. 'Minister' is one of his favourite words, both as verb and as noun. He writes a few lines later of the glorious atom that did 'issuing forth proceed / And Minister in many a Glorious a Deed' (III, p. 353). As the 'Profit' of the individual atom is to 'Minister / Unto the World', so the individual soul should emulate this model by remaining active in its pursuit of knowledge and involvement with 'All Things'. The closing lines of the quotation stress this message, with their insistence that selfhood – the experience of 'I' – is 'More glorious made by coming to the view / Of all that see'. The slight ambiguity around 'all that see', which refers to all things that have perception but also gestures at the soul's comprehensive vision, demonstrates the ideal interconnectedness between human being and all creation. To heighten this further, the 'Pure' and 'Activ' state is contrasted by the trapping of the 'I', twice, in the isolation of the final line ('I Stood alone, tho in the Sun I were').

The focus of the 'Atom' poems in the *Commentaries* is on the particle as an example for the ideal Christian soul. Traherne makes this plain from the structure of his verses, which pursue various activities of the particle in turn, pausing at regular intervals to reflect on the similarities between atom and soul – and what contemplation of the former can do for the latter. Near the beginning of the commentary, he writes of the atom that, 'Like the Soul, it is Empty in it self, and capable of all: but capable of all in an inferior maner' (III, p. 344). Souls, he reminds us in the second poem, are 'Atoms too, and simple ones' (III, p. 362). Traherne explores the capacious 'Simplicitie' of the soul most effectively by meditating on the state of childhood innocence, one of the most distinctive themes of his writings and especially his poetry. For Traherne, as Edmund Newey has explained, 'the child is an image by which the whole story of the human race, from creation to eschatological glory, can be viewed': the experience of infancy offers insight not only into the lost state of prelapsarian bliss, but also, importantly, into the continued ability of the soul to feel joy in 'All Things'.[64] This is explored movingly in the Dobell lyric 'My Spirit':

> My Naked Simple Life was I.
> That Act so Strongly Shind
> Upon the Earth, the Sea, the Skie,
> That was the Substance of my Mind.
> The Sence it self was I.
> I felt no Dross nor Matter in my Soul,
> No Brims nor Borders, such as in a Bowl
> We see, My Essence was Capacitie.
> That felt all Things,
> The Thought that Springs
> Therfrom's it self. It hath no other Wings
> To Spread abroad, nor Eys to see,
> Nor Hands Distinct to feel,
> Nor Knees to Kneel:
> But being Simple like the Deitie
> In its own Centre is a Sphere
> Not shut up here, but evry where. (VI, p. 26)

Traherne does not confine the state of pure being in 'My Spirit' to the infantile perspective but affirms throughout his works that the human is ever able, like the subject of his poem, to experience the world as 'Substance' of the mind and to understand 'I' as synonymous with 'Sence it self'. He portrays the consciousness of the act of being as at one with the universe and defined by its interaction with it. Although the very act of meditation and writing on the soul seems to require a distanced observation of the soul's investigations, recognition of this would be counterproductive according to his philosophy. Writing on Traherne, Denise Inge explains that in 'moments of self-consciousness, in which the self is a subject separate from its object, infinity is a lack. In moments of communion

---

[64] Edmund Newey, '"God Made Man Greater When He Made Him Less": Traherne's Iconic Child', *Literature and Theology*, 24.3 (2010), 227–41 (229).

between a subject and its object, it is experienced as *capacity*' (italics mine).[65] As 'My Spirit' demonstrates, it is the divine duty of a soul to connect with other entities, as its essence is 'Capacitie. / That felt all Things'.

Writing on 'My Spirit', James Balakier comments that the 'felicitous "Self" is filled with joy, overflowing with energy, enlarged by a heightened sense of "Me"'.[66] This emphasis on a powerful selfhood, defined by its engagement with the cosmos, echoes the spiritual volatility and capaciousness of the first atom poem. In 'My Spirit', a sense of 'overflowing' is heightened by the poetic form, specifically the varying numbers of beats between Traherne's iambic lines. Beginning in iambic tetrameter, the second and fourth lines contract to trimeter before the poem opens to the warm breadth of pentameter, which mirrors the 'Essence' of the soul by creating the impression both of great fulness and immense space. Traherne describes his experience of feeling 'all Things' initially through negation ('I felt no Dross nor Matter in my Soul, / No Brims nor Borders, such as in a Bowl / We see'.) Despite the verbal emphasis on what he did *not* feel, the expansion to iambic pentameter, combined with the increased listing of nouns, develops a positive, active impression of the soul's ability to connect with all objects. The enjambment following 'Bowl' is aptly placed to enact the sense of a self without borders. It is significant here (and momentarily puzzling) that Traherne addresses the soul as 'Capacitie' but stresses that it is *not* like a bowl. There is no limit to the soul's containment, as he suggests when claiming 'That Act' was 'the Substance of My Mind'. The 'Act' referenced here is the divine power, that 'Simple Absolut Act' (*Commentaries*, II, p. 171) manifested through creation, which the enquiring, innocent soul is able to absorb and experience as 'Sence it self'.

This mystery of spiritual access and understanding, so unfathomable and prone to gnomic complication in Traherne's prose, is brought to human experience through the swift movements of his poetry. He applies iambic dimeter to explain his essence 'That

---

[65] Denise Inge, *Wanting Like a God: Desire and Freedom in Thomas Traherne* (London, 2009), p. 62.
[66] Balakier, 'Felicitous Perception', p. 60.

felt all Things, / The Thought that Springs / Therfrom's it self'. The poem physically springs from pentameter to two lines of dimeter and back again, motions of contraction and dilation that recreate the act of experiencing 'all Things'. Through these metrical transitions, the reader can grasp the paradox of knowledge that is both universal and self-originating: as in the capacious soul, so in the space of the poem. Traherne's use of 'Springs' suggests the joyful leap of active contemplation, but also recalls his habitual use of Neoplatonic water imagery to describe the source of divine act as a 'fountain'. As Traherne continues to develop the activity of his ideal, energetic 'Spirit', strong parallels emerge between the vocabulary of the insatiable soul and the language of the atom. There is resonance between his indivisible atom and the Cusanian conclusion to the first stanza, that the soul, 'Simple like the Deitie / In its own Centre is a Sphere / Not shut up here, but evry where'. Across Traherne's works, a common language arises between references to God, soul and atom, all of which contain an 'unsearchable Abyss of wonders' (*Commentaries*, III, p. 333). All things are connected but, as Traherne shows in his discussion of the atom as a 'means', there is a hierarchy of accessibility in this cosmos. Active contemplation in and of the soul permits intuitive understanding of the deity, to the extent that, in realising the soul's 'Simplicitie' in 'My Spirit', the poet can discover the power of its divine source. To begin this process of realisation, comprehension of the soul takes form by borrowing vocabulary from Traherne's atom. The soul in 'My Spirit' has the power to:

> Dilate it self even in an Instant, and
> Like an Indivisible Centre Stand
> At once Surrounding all Eternitie.
> Twas not a Sphere
> Yet did appear
> One infinit. Twas somwhat evry where.
> And tho it had a Power to see
> Far more, yet still it shind
> And was a Mind
> Exerted for it saw Infinitie
> Twas not a Sphere, but twas a Power
> Invisible, and yet a Bower. (VI, p. 29)

The sixth stanza makes explicit the hierarchy Traherne draws between soul and atom, that the latter, 'Like the Soul ... is capable of all: but capable of all in an inferior maner' (III, p. 344). His superior infant soul 'saw Infinitie' but 'had a Power to see / Far more'. Once again, the metrical movements of the verse enact the soul's wondrous ability to contract its focus and to 'Dilate', encompassing not merely creation, but 'Eternitie'. An added syllable to the first quoted line gives energy to this dilation, before the poem shrinks back to dimeter to emphasise the human capability of this wonder ('Twas not a Sphere ... / And was a Mind'). Traherne's soul is superior to his atom, but to recognise its properties he applies details from his atomic philosophy. Soul is '*not* a Sphere' (italics mine) but appears like one, recalling the claim in the *Commentaries* that the atom, similarly indivisible, has no shape but bears the divine emblem of the sphere as its 'Conceivable Figure'. The soul is 'Like an Indivisible Centre', a similitude that evokes the indivisible 'Act' of the divine but also the indivisible centres of atoms, which, Traherne explains in the *Commentaries*, are likewise 'somwhat evry where'. By returning frequently in his atom poems to the model of the indivisible's journey through creation, he demonstrates how the particle illuminates the soul's insatiability as 'One single Atom runs a curious Race, / And is always the same in evry place' (III, p. 361).

His depictions of atomic volatility, combined with the ability of his soul in 'My Spirit' to 'Dilate it self ... in an Instant', reveal the influence of the hermetic corpus – the chief source of inspiration behind Traherne's recognition of 'All in All', the immanent state of being and comprehension applicable to both souls and atoms.[67] He

---

[67] These mystical writings on mind, nature and the cosmos were considered the work of the mythical Hermes Trismegistus, supposedly a combined representation of the Greek god Hermes and the Egyptian god Thoth. In 1614 the classical scholar Isaac Casaubon denied the authorship of the ancient Trismegistus, suggesting that the mystical writings were 'the figment of a half-Christian forger' rather than the work of 'a contemporary of Moses'. After Casaubon's denouncement of ancient authorship the texts did seem to gravitate further towards a 'popular' audience, but the hermetic corpus still captivated scholarly philosophical and theological thought: most notably, in the writings of the Cambridge Platonists and Thomas Traherne. See *Hermetica: The Greek Corpus Hermeticum and the*

is moved by the hermetic insistence that the soul inspires infinite exploration and understanding, to the extent that he requoted a passage from 'The Tenth Book of Hermes Trismegistus' multiple times across his works: 'Increase thy self to an immeasurable Greatness, leaping beyond every Body, and transcending all Time, become ETERNITY; And thou shalt understand GOD'.[68] The statement encourages the soul on its pursuit of greatness, declaring that its desired reunion with the divine 'Act' is achievable. Hermetic movements of dilation and contraction are applicable equally to Traherne's atom, which, like his soul, must be occupied in act to achieve its divine potential. In *The Kingdom of God*, he writes of atoms that 'In evry Sand, in evry Grain of Dust there is an Infinit Number, yet every one of these is a Temple of his Omnipresence' and 'a means of our Happiness' (I, p. 343). As the repeated reference to 'means' here suggests, the atom is available to aid the understanding on its enquiries. Each particle is 'a Temple' of God's universality, implying that it is open to the soul as a place for worship and reunion with the divine act – a site for 'Being with the Being', as Traherne describes pure, active contemplation in 'My Spirit' (VI, p. 27).

Latin Asclepius in a New English Translation, trans. Brian P. Copenhaver (Cambridge, 1992), p. 1. Traherne quotes from the *Hermetica* avidly in his commonplace book and throughout his published writings, taking as his source an English translation produced by John Everard in 1649. John Everard, *The divine pymander of Hermes Mercurius Trismegistus, in XVII books* (London, 1649).

[68] One of such quotations from his commonplace book, a description of infinite spiritual ability taken from 'The Tenth Book of Hermes Trismegistus', was reproduced both in *The Kingdom of God* and in *Christian Ethicks*, the latter from which I quote:

And judge of this by thy self. Command thy Soul to go into India, and sooner than thou canst bid it, it will be there. Bid it pass over the Ocean, and suddenly it will be there ... Increase thy self to an immeasurable Greatness, leaping beyond every Body, and transcending all Time, become ETERNITY; And thou shalt understand GOD. (p. 226)

See also Traherne's commonplace book, fol. 23v, and *The Kingdom of God*, I, p. 464. For the original, see Everard, *Divine Pymander*, pp. 153–54.

This state of spiritual bliss is captured most faithfully and, in a literal sense, movingly, by the flights of his lyric poetry. Poems, like the atom and like the soul, are equipped to recreate the hermetic movements of dilation and contraction and to transfer theological obscurities into the realm of active experience. Through their dynamic motion, Traherne's poems form the ideal vehicle for active contemplation of the soul and the atom, the intermediary, material indivisible key to the comprehension of spiritual capacity. While atoms are characterised as 'passiv' in Traherne's philosophical prose (*The Kingdom of God*, I, p. 342), the poems in the *Commentaries* recreate an indivisible that is active in its intimate relationship with the immortal soul. I wish to conclude by returning to the poetry with which I opened this chapter, Traherne's claim at the end of the commentary that 'Souls are Atoms too', yet 'for these Souls all Atoms were prepard'. The second poem continues, referring to the powers of the soul in comparison with 'The Rest' (atoms):

> But then their Essence they another Way
> Unknown to Atoms like them selvs enjoy
> And being Nothing, but a Power to see
> And Good for Nothing but Felicitie
> Their Nature and their Motion is to lov
> Their Being to behold their Life above.
> The Rest are Good for Nothing but to serv
> These Good for Nothing if from Bliss they swerv. (III, p. 363)

The 'Bliss' experienced by the soul is in a focused self-sufficiency, while atoms are made to 'serv'. Traherne emphasises the need for spiritual focus in his reference to the 'swerv' of inattentive souls. Recalling the Lucretian account of the 'swerve' to describe the motion of atoms in chaos, the metaphor accentuates the demarcating prerogative of the soul to remain concentrated in self-concentration. As they are 'Nothing' but the power 'to see' and, as was quoted earlier, 'very Nothings they appear', it seems that souls come to 'appear' when they fulfil the act 'to see'. This circularity is heightened in the repetition of 'being' as verb and 'Being' as noun. While the status of 'being Nothing' is oxymoronic, the tension is eased with the flexible possibilities of the third couplet. The lines 'Their Nature and their Motion is to lov / Their Being to behold their Life above' could be read separately, with the sense that love is in the nature of souls but their very existence is defined through the act of beholding eternal life. Alternatively, the suggestive enjambment after 'lov' could be an instruction that they are required to 'lov / Their Being' to understand the glories of celestial life. Once again, the movements of Traherne's poetic form enable circular, mutually enforcing interpretations, protecting the multiplicity of meaning inherent to exploring a divine concept. We are told our existence and the very continuation of love relies on our ability to view the 'Life above', but this perspective in turn depends on the act of loving our being. It is in the act of seeing, of experiencing and possessing knowledge, that souls fill the void and come to possess a 'Being' that shows them the 'Way' to spiritual ecstasy.

The atom gives the soul directions to this advanced state. The familiarity Traherne establishes between atom and soul corroborates his favourite hermetic theories. He writes in the earlier poem from the commentary:

> Nor is it any Wonder that the King
> Of Glory should creat so Great a Thing,
> In evry Soul; since in an Atom we
> Such Endless Depths of Alsufficiency
> Behold, and find so may [sic] thousand Parts
> Shut up in one, my Miracles of Arts. (III, p. 356)

The shift from first person plural to singular, made more emphatic by the possessive, stands out in the poem. With 'my Miracles of Arts' Traherne may have in mind a work by the thirteenth-century friar and alchemist Roger Bacon, the *Discovery of the Miracles of Art, Nature, and Magick*, which received its first published English translation in 1659. Traherne's philosophy of the soul had much in common with Bacon's. The earlier philosopher declared that 'there is not any thing, whether in divine or outward matters too difficult for my faith' and that the 'rational soul is not impeded in its operations, unlesse by the Manicles of ignorance', statements that would not be out of place in Traherne's *Commentaries*.[69]

---

[69] Roger Bacon, *Frier Bacon his discovery of the miracles of art, nature, and magick* (London, 1659), p. 25; p. 16. Bacon continues, citing Aristotle and the Neoplatonic philosopher and physician Isaac Israeli:

Here again, the unfathomable becomes fathomable as Traherne invites us to 'Behold' the 'Endless Depths' within the atom. His vocabulary recalls the intuitive knowing, described earlier in the commentary, of feeling 'in an Atom (with our Minds)' – of knowledge gained through empathy, through recognition of the affinity between the interior of the atom and the infinite capaciousness of the soul. An obvious problem arises with the claim that we may find 'so ma[n]y thousand Parts / Shut up' within the atom. An atom cannot, of course, be divided into physical parts, but Traherne's approach here is both metaphysical and poetical. The emphasis is not on the atom in isolation, but on the qualities that transpire from the soul's interaction with material first principles. As a means to knowledge and the first and last of created entities, the atom contracts secrets for the soul to seek; the 'many thousand Parts' refer to its role as a vessel for divine knowledge and act. If, as previously explored in Traherne's poem, an atom has experienced material being from 'Mirie Fen' to glorious sun, it has accumulated numerous understandings of divine activity to share with the curious soul. Poetic form similarly stores copious information for the ensouled reader. In this extract, Traherne's couplets enact the constancy and continuity of the atom even as they also stress the expansion of divine desire. The poetry bursts from the closure of its rhyme-scheme, refusing to contain its subjects and thus demonstrating the unlimited power of act: the important location of the 'Great … Thing' is revealed to be 'evry Soul' at the beginning of the next paired lines; likewise, the main verb of the following idea, 'Behold', is pushed to the start of the new couplet. Here and elsewhere, the structures of Traherne's poetic spaces present numerous possibilities for interpretation through their embracement of paradox, metrical variety and wordplay.

'Twas excellently said of Isaac (in lib. de Febribus,) The rational soul is not impeded in its operations, unlesse by the Manicles of ignorance. And Aristotle is of opinion, (in lib. secret.) That a clear and strong intellect, being impregnated by the influences of divine Virtue, may attain to any thing which is necessary. And in 3[d] Meteor, he saith, There is no influence or power, but from God. In the Conclusion of his Ethicks, There is no Virtue, whether Moral or Natural without divine influence. (p. 16)

Traherne argues that a complete understanding of 'ALL THINGS' is possible not only in divine omniscience but also from the position of embodied mortality – and not only possible, but essential. Through Traherne's explorations of the natural world, soul and Godhead, the distinctions between corporeal and spiritual states of being – mortal and immortal, ecstatic and 'instatic' (to borrow the term from Robert Watson) – become indistinguishable and 'The Kingdom of God' opens to terrestrial enquiry.[70] For Traherne, accepting and comprehending the atom was a necessary part of trusting in the mysteries of the Christian faith. The atom, 'a great Miracle in a litle Room', provides a 'means' for the soul in its pursuit of active contemplation. Themselves the 'Mobile vigorous Active Spirits of the Univers', atoms lead the way for the soul on its mission to discover all objects in the universe, abstract and physical (III, p. 342). Traherne's endeavour to verbalise this mystical bond creates a poetics of the atom in his writings. His poems do more than preserve the active relationship between atom and soul: they inspire it. The subtle and volatile movements of his lyrics complement and enact the motions of the indivisibles – atoms and souls – he locates as omnipresent throughout creation. To conceive of a mystery, it is necessary for it to take form. Poetry both accommodates and activates Traherne's intuitive explorations of the divinity within and beyond the self. Ultimately, the hundreds of pages of Traherne's philosophical writings could contract to a single observation, fathomed in poetry: that 'Souls are Atoms too, and simple ones'.

---

[70] Robert N. Watson, *Back to Nature: The Green and the Real in the Late Renaissance* (Philadelphia, 2007), p. 297.

Chapter 3

WORLD-MAKING AND WORLD-BREAKING: THE ATOM POEMS OF MARGARET CAVENDISH AND HESTER PULTER

> ... by other atoms thrust and hurled
> We give a being to another world.
> Hester Pulter, Poem 58
>
> Small atoms of themselves a world may make ...
> Margaret Cavendish, 'A World Made by Atoms'[1]

Although the contemporaries Margaret Cavendish and Hester Pulter might seem to share general affinities – they were both female, poets and Royalist aristocrats – a comparison of their lives and works finds significant differences. While the writings

---

[1] Hester Pulter, 'Poem 58', in *Poems, Emblems, and the Unfortunate Florinda*, ed. Alice Eardley (Toronto, 2014), p. 169. All quotations from Pulter's poetry will be taken from this edition and cited in the text by page and line numbers. For an online edition of Pulter's works see *The Pulter Project*, ed. Wendy Wall and Leah Knight <http://pulterproject.northwestern.edu/> [accessed 26 Feb. 2021]. Margaret Cavendish, *Poems and Fancies*, quoted from *Margaret Cavendish's Poems and Fancies: A Digital Critical Edition*, ed. Liza Blake <http://library2.utm.utoronto.ca/poemsandfancies/> [accessed 26 Feb. 2021]. For 'A World Made by Atoms', see <http://library2.utm.utoronto.ca/poemsandfancies/2017/06/09/a-world-made-by-atoms/> [accessed 26 Feb. 2021]. All further quotations from *Poems*

of 'Margaret the First' have long since graced library shelves – including the collections to which she personally bequeathed copies in the 1660s – it seems unlikely that Pulter ever intended her 'Fancies' to see print.[2] In contrast to Cavendish's loud self-fashioning through incessant publication and autobiographical statement, Pulter appears to have lived rather reclusively, raising fifteen children from the family seat in Hertfordshire and compiling some 120 poems, highly devotional in content, into a private manuscript over the course of twenty years.[3] Her poems, an assortment of

---

and Fancies will be taken from this edition and will be cited in the text by line number(s). Blake presents a collated edition of the three versions of *Poems and Fancies* published in Cavendish's lifetime: *Poems, and fancies* (London, 1653); *Poems, and phancies* (London, 1664); and *Poems, or, Several fancies in verse, with the Animal Parliament in prose* (London, 1668). An immense contribution to scholarship, the digital edition is accompanied by detailed notes explaining textual variants between these three versions. Blake explains the editorial process in the textual introduction:

> we found that any choice among editions lost what we found to be the most interesting aspect of her poetry: the fact that it exists in (at least) two very different versions, each of which has its own apparent advantages and disadvantages. We decided, then, that no matter what editorial strategy we ultimately pursued, we wanted an edition that had full textual notes, so that readers could use our edition to understand the poems in their full and complicated textual history, and could themselves grapple with the numerous changes she made as she revised her poems over 15 years.

See Blake, 'Textual and Editorial Introduction' in *Margaret Cavendish's Poems and Fancies*, <http://library2.utm.utoronto.ca/poemsandfancies/textual-and-editorial-introduction/> [accessed 26 Feb. 2021]. I discuss Cavendish's revisions to her text later in this chapter.

[2] Cavendish herself contacted publishers with her manuscripts in addition to donating many of her works to Oxford and Cambridge college libraries. For more information on her self-promotion in publication see Katie Whitaker, *Mad Madge: Margaret Cavendish, Duchess of Newcastle, Royalist, Writer and Romantic* (London, 2003), pp. 158–59, 192 and 312–14. For the title 'Margaret the First', see Cavendish's preface 'To the Reader' in *The description of a new world, called The Blazing World* (London, 1666), sig. b*2r; see also Douglas Grant, *Margaret the First: A Biography of Margaret Cavendish Duchess of Newcastle 1623–1673* (Toronto, 1957).

[3] For further details about what can be surmised of Pulter's life, see *Poems*, ed. Eardley, pp. 13–21 and 'The Poet' in *The Pulter Project*, <https://

meditations on miscellaneous subjects, are addressed to the select audience of herself, her daughters and God – a figure that, incidentally, the natural philosopher Cavendish rarely mentions.

Yet, despite their differences in literary intent and ambition, the writers have a notable common interest: they both wrote poems about atoms. Cavendish and Pulter discovered imaginative potential in the atom and employed poetry as the medium for teasing out its productive paradoxes. In their writings, the 'atom poem' materialised as a recognisable form: it could stand alone or contribute to a general theme, building with other poems in sequence; it experimented with the making and unmaking of different things; explored hypotheses; contracted different ideas, relationships or experiences into one unit, or potentially 'indivisible' point. Their poems merged with the atom in ways that were both mechanical and thematic. Not merely the contents but also the structures of their verses were held together by the fundamental principles of atomic generation and corruption, resting on the paradox that entities so close to nothingness may 'a new world create'.[4]

It is the creation of worlds – their making, breaking and renewing – that emerges as the most prominent, significant focus of all Pulter and Cavendish's atom poems. New worlds promised and created through writing are promised and as good as made by atoms, which underlie every creative act, including the material letters of the text itself and the fate of the respective authors.[5] No mere quantitative components of physical matter, atoms provided a means for constructing (in Cavendish's case) and deconstructing (in Pulter's) authorial and spiritual self. Atomic volatility is understood not as threatening but as a liberating force within these poetic

---

pulterproject.northwestern.edu/about-hester-pulter-and-the-manuscript.html> [accessed 26 Feb. 2021].

[4] Margaret Cavendish, 'A World Made by Atoms', line 17.

[5] Lucretius famously punned on 'elementa' in drawing a close comparison between atoms and letters. Alessandro Schiesaro has commented on the deeper significance of this: 'there can be no analogue which is ontologically independent of the atomic reality it supposedly represents'. Alessandro Schiesaro, 'The Palingenesis of *De Rerum Natura*', *Proceedings of the Cambridge Philological Society*, 40 (1994), 81–107 (87).

spaces. It is to the benefit of individuals – human beings, ideas and bodies – that they are fragmentary, imperfectly formed, or liable to change. In poetic forms that are analogous to these unfixed states of being, these writers discover liberty of faith, expression and creative potential in that which can be done and undone. Pulter literally loses herself within the atomic environments of her poems, exercising the freedom of her verse as she takes intimate knowledge of the universe as her subject. Cavendish associates the movement of atoms in her poetry, and the figures they make, with the measurement of her authorial success.

Theirs was no strict classical atomism, derived solely via Epicurus and Lucretius; nor were their creative theories in debt to Descartes's inert, mechanised corpuscles. Cavendish's atoms assumed Epicurean properties of shape and volatility, but were endowed with vitalist impulses, even agency and personality, in their movements.[6] Pulter kept the atomist concept of material indivisibility but combined it with the strong influence of Aristotelian *minima*, minimal – though not necessarily divisible – particles of matter descended from medieval Scholastic philosophy and alchemy. Nor would this influence have escaped Cavendish. As recent work by historians of science (notably Antonio Clericuzio, William R. Newman and John Henry) has shown, the majority of western 'atomic' philosophies, pre-Gassendi, owed more to Julius Caesar Scaliger's reinterpretation of Aristotelian *minima* as 'indivisibles' than to Lucretius and Epicureanism.[7] Daniel Sennert,

---

[6] Cavendish was highly critical of Descartes's philosophy, dedicating a section of her *Philosophical Letters* to a detailed critique of his physics. See *Philosophical letters: or, Modest reflections upon some opinions in natural philosophy, maintained by several famous and learned authors of this age, expressed by way of letters* (London, 1664). See also Stephen Clucas, '"A double Perception in All Creatures": Margaret Cavendish's *Philosophical Letters* and Seventeenth-Century Natural Philosophy', in *God and Nature in the Thought of Margaret Cavendish*, ed. Brandie R. Siegfried and Lisa T. Sarasohn (London, 2014), pp. 121–39.

[7] See, for example, Antonio Clericuzio, *Elements, Principles and Corpuscles: A Study of Atomism and Chemistry in the Seventeenth Century* (Dordrecht, 2000); William R. Newman, *Atoms and Alchemy: Chymistry & the Experimental Origins of the Scientific Revolution* (Chicago, 2006); John Henry,

Walter Warner, Thomas Harriot – the work of whom Cavendish would have been introduced to by her brother-in-law, Charles – and Francis Bacon all combined atomism with Aristotelian theories of forms and chemical properties.[8] Slightly later in the century, Kenelm Digby's popular *Two Treatises* was published and read by a wide audience which almost certainly included Cavendish, and quite possibly Pulter.[9] While Digby rejected Aristotelian notions of form and prime matter, his (potentially divisible) 'atoms' descended directly from the *minima naturalia* tradition. For Pulter especially, this tradition put atoms into contact with chemical vitalism, thus permitting the mingling of atomic re-formation with themes of life, creation and rebirth.

Cavendish and Pulter's atoms were intimately associated with spirit, character and new life. In this chapter, I explore the nuances of the relationship between the atom and the writers' poetical, philosophical and theological ambitions. The key word for Cavendish is 'reinvention', the construction of self and authorship through atomic collaboration and formation; for Pulter, it is 'resurrection' – not through constructive creation, but via the negative stroke of dissolution. Whether perceived through the lens of utmost humility (Pulter) or immense authorial potential (Cavendish), both directions of atomic movement, corruption and generation, bore the power to liberate: to enable freedom of expression and intimate knowledge of one's subject, be it natural

---

'Atomism and Eschatology: Catholicism and Natural Philosophy in the Interregnum', *The British Journal for the History of Science* 15.3 (1982), 211–39.

[8] Charles Cavendish worked closely with Harriot's work and copied one of his manuscripts. For further information, see Stephen Clucas, 'The Atomism of the Cavendish Circle: A Reappraisal', *The Seventeenth Century*, 9.2 (1994), 247–73.

[9] Kenelm Digby, *Two treatises in the one of which the nature of bodies, in the other, the nature of mans soule is looked into in way of discovery of the immortality of reasonable soules* (London, 1644). Digby's natural philosophical work was exceptionally popular: the text had undergone five English editions by 1669, plus two editions in Latin which facilitated its reception on the continent. For further details, see Paul S. Macdonald, *Kenelm Digby's Two Treatises* (Gresham, 2013), p. 31.

philosophical or theological. The miraculous atom was next to nothing, but simultaneously in close contact with all things, divine and natural, due to its unique position at the origins of creation. Cavendish understood this perfectly when she quipped that her poems on atoms might either make her 'a world, or nothing' ('To Natural Philosophers', *Poems and Fancies*).

Margaret Cavendish, Duchess of Newcastle cuts a rather unusual figure in early modern literary history. Only the second Englishwoman to be published under her own name during her lifetime, she wrote specifically for print, and published thirteen books in a diversity of genres including poetry, prose romance, drama, biography, the epistolary form and the philosophical essay.[10] The variety of the forms in which she experimented – doubled with the fervency of her self-promotion – fittingly represents her constant textual and natural philosophical preoccupation with reinvention. From the beginning to the end of her literary life, studies in natural philosophy formed the most consistent topic of her writings. Across her works, Cavendish defends the originality of her scientific theories, the central themes of which are the materiality of all created things, including thoughts; the possibility of infinity in body (she maintains that nature is both infinite and corporeal); the innate subjectivity of ideas and the interior freedom of perception within the beholder, regardless of external contact with other objects. Her rationalist philosophy promotes a form of vitalist materialism, in which she understands all matter to have sense perception and to be self-moving.[11] It is Cavendish's response and contribution

---

[10] The first woman to be published in England (without her permission) by name was Anne Bradstreet, with *The tenth muse lately sprung up in America* (London, 1650).

[11] It is necessary to distinguish between some terms so as not to provoke confusion when exploring Cavendish's philosophy. Cavendish conceives of all things in the natural world as 'material', which she considers synonymous with 'corporeal'. Though not all are solids, she would classify a table, a cloud, a thought and the human soul all as material substances. They are constituents of nature and conform to the characteristics that distinguish parts of that infinite body: they are 'self-moving and self-knowing' (and therefore in possession of the agency necessary to distinguish them). Because she argues that it would be impossible to conceive of motion or

to seventeenth-century science that has attracted the greatest amount of critical attention on her work in recent years.[12] She was just as ambitious, however, as an author of poetical 'Fancies', and continued to produce what later readers would class as 'fictional' works for the entirety of her working life. These texts, among them *Natures Pictures* (1656) and the well-known *The Blazing World* (1666), offered imaginative commentaries on natural philosophical ideas often paralleled in Cavendish's essays.[13] The addition of works composed in forms occupying ambiguous positions in-between literary genres – such as the *CCXI Sociable Letters* and *Orations upon Divers Subjects* – further suggests that it would be misleading to sever the philosophical from the poetical in Cavendish's work, or to prioritise one discipline over the other by adhering too closely to the author's prefatory defences.[14] Cavendish's dual authorial

---

form without body, Cavendish classifies all beings and actions as corporeal, even if it would be more typical to perceive them as abstractions. She defends this with the greatest clarity in *Observations upon Experimental Philosophy*:

Nature is purely corporeal and material, and there is nothing that belongs to, or is a part of nature, which is not corporeal; so that natural and material, or corporeal, are one and the same … Neither is there any such thing as an incorporeal motion; for all actions of nature are corporeal, being natural; and there can no abstraction be made of motion or figure, from matter or body, but they are inseparably one thing.

See *Observations upon Experimental Philosophy*, ed. Eileen O' Neill (Cambridge, 1998), p. 137. All further quotations from this text will be taken from this edition and cited in the text by page number. Cavendish also writes in the *Philosophical Letters*: 'Motions, Forms, Thoughts, Ideas, Conceptions, Sympathies, Antipathies, Accidents, Qualities, as also natural Life, and Soul, are all Material' (p. 12).

[12] See, for example, Lisa T. Sarasohn, *The Natural Philosophy of Margaret Cavendish: Reason and Fancy during the Scientific Revolution* (Baltimore, 2010), Karen Detlefsen: 'Atomism, Monism, and Causation in the Natural Philosophy of Margaret Cavendish', *Oxford Studies in Early Modern Philosophy*, 3 (2006), 199–240 and Richard Nate, '"Plain and Vulgarly Express'd": Margaret Cavendish and the Discourse of the New Science', *Rhetorica: A Journal of the History of Rhetoric*, 19.4 (2001), 403–17.

[13] *Natures pictures drawn by fancies pencil to the life* (London, 1656).

[14] *CCXI sociable letters* (London, 1664) and *Orations of divers sorts, accommodated to divers places* (London, 1662).

approaches of poetical 'Fancy' and philosophical 'Reason' were consistently, and necessarily, entwined, and her 'atom poems' – the opening of her first publication, *Poems and Fancies* (1653), which endorsed her initial support for and unique take on the atomic hypothesis – were the origin of this fused relationship.

Hester Pulter was no philosophical atomist, but her 'atom poems', though fewer in number than Cavendish's, are rich in philosophical complexity.[15] She gathered poems into a manuscript between 1645 and 1665, the bulk of which appear to have been composed during the Civil War and Interregnum – a span of time in which the Pulters withdrew from public life and settled into a new family home at Broadfield, Hertfordshire.[16] Her poems do not appear to have been circulated, and several comments in the manuscript indicate that she wrote during her periods of confinement.[17] In spite of the intellectual isolation suggested by her social and geographical circumstances, Pulter wrote poems that engaged directly and knowingly with current public affairs, theological, political and scientific. Her work reveals a tendency to relate her readings in divinity and natural philosophy to the everyday objects of her experiences: the flowers in her garden; a swarming anthill; the beauty of her children. But Pulter also turns to the tragedies of war and to personal grief. Between 1624 and 1648 she gave birth to fifteen children, only two of whom were to outlive her.[18] Frequent bouts of sickness haunt her poems, as do plaints of suffering and melancholy reflections on ageing. These reflections are frank and perhaps symptomatic of the secluded context of her writing. Unlike Cavendish, Pulter may not have been writing with any kind of external audience in mind, but in response to her own fears, desires and observations.[19]

---

[15] Only five poems and two emblems from Pulter's manuscript mention atoms directly, but, as I shall show later in this chapter, an atomic vocabulary nonetheless characterises much of Pulter's verse.

[16] Pulter's manuscript was rediscovered at the Leeds Brotherton Library in 1996, and is Leeds, Brotherton Collection, MS Lt q 32.

[17] See Mark Robson, 'Pulter, Lady Hester (1595/6–1678)', in *Oxford Dictionary of National Biography* <http://www.oxforddnb.com/view/article/68094> [accessed 23 Aug. 2013].

[18] *Poems*, ed. Eardley, p. 16.

[19] The editors of The Pulter Project note, however, that though Pulter's writings 'seemingly lay unread ... for over 250 years', her poems were

Cavendish's contributions to seventeenth-century natural philosophy have been well-noted, if not entirely satisfactorily so, but Pulter's significant interest in the contemporary physical sciences remains largely unexplored. Though predominantly devotional in content, her poems jump between intellectual and experimental observations, prompted by her revolving turns as naturalist, astronomer, metaphysician, political theorist and elegist. Sarah Hutton has argued convincingly that Pulter appears to have been aware of the writings of Donne and of Marvell, from where she may have discovered many of her natural philosophical ideas.[20] Yet Pulter goes beyond Donne's cold 'atomies', which exist insofar as they are symptomatic of a 'crumbling' world order – when evoked, Pulter's 'indivisibles' pose a sincere spiritual purpose. Their appearance in her poems strengthens a faith in the theory that the world, and all living beings within it, is formed from atoms, and that these atoms are indispensable components for the realisation of eternal life. This is not to say that she writes in order to argue the atomic hypothesis: as Hutton continues, it would be misleading to claim that Pulter wrote '"scientific poetry" in the sense that the purpose of her poems was to convey scientific theory or scientific knowledge in poetic form'.[21] Her poems almost take the corpuscular structure of matter for granted in their explorations of natural and spiritual mysteries. Atoms, worlds and souls rub shoulders in her poetry, resulting in no simple exercise of devotional humility. In Poem 58, Pulter vows that, after death and dissolution, she 'shall surround this spacious universe' (p. 169, 4): the spacious vowels of the statement amplify the promise that the atomised/resurrected Pulter will not only pervade the extended universe but enclose it, tracing its circumference with confident ease.

'consciously gathered neatly for presentation' in the manuscript. See <https://pulterproject.northwestern.edu/about-hester-pulter-and-the-manuscript.html#manuscript> [accessed 26 Feb. 2021].

[20] Sarah Hutton, 'Hester Pulter (c. 1596–1678): A Woman Poet and the New Astronomy', *Études Épistémè*, 14 (2008), 1–19 (3). Louisa Hall has also written compellingly of Pulter's knowledge of and 'faith in the Copernican system'. Louisa Hall, 'Hester Pulter's Brave New Worlds', in *Immortality and the Body in the Age of Milton*, ed. John Rumrich and Stephen M. Fallon (Cambridge, 2018), pp. 171–86 (p. 171).

[21] Hutton, 'A Woman Poet', p. 4.

Her promise of empowerment is Cavendish-like in its assertiveness. In the preface 'To Natural Philosophers' from *Poems and Fancies*, Cavendish makes the following famous declaration of ambition, associating firstly her compositions and then herself with the atom and its movements:

> my desire that they should please the readers is as big as the world they make, and my fears are of the same bulk. Yet my hopes fall to a single atom again, and so shall I remain an unsettled atom, or a confused heap, till I hear my censure. If I be praised it fixes them, but if I am condemned I shall be annihilated to nothing. But my ambition is such, that I would either be a world, or nothing.

The 'they' under discussion here are her atoms and her poems, both respectively and compositely. Cavendish's poems activate worlds within worlds. Her desire to 'please' may be as big as the 'world', but her 'hopes' are atoms – what may seem modest humility at first glance doubles as extravagant confidence, as the poems themselves replace the world that not only inspired but provided the raw materials for their creation. Despite Cavendish's message here, there is no simple opposition between 'world' and 'nothingness' in her thought. Her hopes not only 'fall to' but rest on single atoms, as, parallel to her texts, she adopts the status of a particle herself ('I remain an unsettled atom... If *I* be praised it fixes *them*', italics mine.) An atom, while no-thing, contributes to the matter of the world and is a world in itself. This is something recognised by both Cavendish and Pulter: to associate one's authorial self with atomic behaviour is at once a careful display of modesty and an assertion of the individual's ability to change the world – even to create anew.

Over the following pages I explore and challenge Pulter and Cavendish's unconventional takes on the atom, questioning its place in their poetry and the role of the 'atom poem'. Both recognised something in the atom that fed into poetical expression and the lyric form. Not unlike the heroic couplet favoured by both writers – a memorable, pithy device through which 'complex issues' could be encapsulated and 'submitted to sense and reason' – meditations on atomic creation opened up major, otherwise unfathomable topics within their poetry.[22] The atom became an object for the imaginative investigation of these subjects. Occupying a space in-between the knowable material world and divine secrets, it offered greater insight into the biggest mysteries of the Christian faith while still preserving their deep ineffability. For Cavendish and Pulter, atoms posed solutions for working through poetical and interpretative challenges, but did so by embracing complications, not by attempting to reduce them. A reading of their readings into the atom unlocks the paradoxes and liberating potential of the indivisible particle.

HESTER PULTER: ATOMS AND EMBLEMS

It would be easy to overlook the significance of the 'poem' as Cavendish and Pulter's preferred literary form for experiment and meditation, but both writers had clearly reflected on the nature of lyric and its place in developing their collections. As her title demonstrates, Cavendish divided 'Poems' and 'Fancies' into related but distinguishable categories. For Pulter, poems occupy a designated section – 'Poems Breathed Forth by the Hadassah Right Honorable' – within a broader assembly of writings entitled 'Hadassah's Chaste Fancies'. She is moreover direct in distinguishing between her 'Poems' and her 'Emblems', likewise the 'Sighs of a Sad Soul', not merely 'Breathed Forth' but *Emblematically* Breathed Forth' (italics mine.)[23] Pulter's writings lack the extensive appeals 'To The Reader' that define Cavendish's characteristic self-promotion, but it is still possible to discern some of her authorial intention through examination of her manuscript's structure, clusters of imagery and bursts of personal insight.[24]

---

[22] William Bowman Piper, *The Heroic Couplet* (Cleveland, OH, 1969), p. 309.
[23] *Poems*, ed. Eardley, p. 185.
[24] Jayne Archer has commented on how it is 'characteristic of Pulter's poetry that certain images and word clusters are reworked', taking the example of Pulter's multiple 'Circle' poems as her study. Jayne Archer, 'A 'Perfect Circle'? Alchemy in the Poetry of Hester Pulter', *Literature Compass*, 2.1 (2005), 1–14 (9).

Pulter's distinction between 'poem' and 'emblem' is key to understanding the nature of her interest in the atom. As a form, the emblem has proved notoriously difficult to define. The now familiar categories of *inscriptio*, *pictura* and *subscriptio*, clearly identifiable in the popular 'emblem books' of sixteenth- and seventeenth-century Europe, cannot be applied with ease to Pulter's 'naked' emblems, which are devoid of pictures and bear no individual titles (unlike the majority of the poems in the opening pages of her collection).[25] Her emblems can be recognised nevertheless as part of a meditative poetical tradition that lifts objects from nature, holds them up to view, and draws specific moral conclusions. The atomistic murmurings of her opening verses narrow down to two appearances over the course of her fifty-two emblematic poems – there is something about the atom which, for Pulter, lends itself to personal reflection and devotion over analysis and symbol. Where she mentions them, her atoms are always in motion or associated with recombinations of form. It is hardly surprising, then, that they feature less frequently over the course of the second half of her collection. 'The true emblematic image stands motionless', writes Thomas Owen Beachcroft in an attempt to define the emblem; Peter Daly pushes this further, to comment on the stark lack of interrelation between the entities that constitute any given emblem, '[t]he natural objects are combined unorganically in one picture; that is, the relationship between the objects is not given in nature.'[26]

It is, however, one of the two emblems in which Pulter addresses 'atoms' that demonstrates her most scientifically informed perspective on corpuscular matter theories. Daly's argument about the 'unorganic' composition of emblematic entities cannot be applied to Pulter, whose emblems, even those on apocryphal or mythological subjects, are consistently based in realistic pastoral settings.

---

[25] For further information on Pulter's emblems, and their break with tradition, see Rachel Dunn, 'Breaking a Tradition: Hester Pulter and the English Emblem Book', *The Seventeenth Century* 30.1 (2015), 55–73.

[26] T. O. Beachcroft, 'Quarles – and the Emblem Habit', *Dublin Review*, 188 (1931), 80–96 (89); Peter Daly, *Literature in the Light of the Emblem: Structural Parallels between the Emblem and Literature in the Sixteenth and Seventeenth Centuries* (Toronto, 1979), p. 90.

The inspiration Pulter gathers from her readings of Ovid and Pliny merges with her close observations of the natural world, forming a rhetoric of specific, itemised description that would not be out of place in works of experimental science. The 'emblem', with its close ties to fable and allegory, was strongly suited to alchemical discourse, where its images might be applied to 'vivify, enrich, clarify, allegorize, mystify or even obfuscate' – the famous illustrations and accompanying *subscriptio* by Robert Fludd and Michael Maier provide notable examples.[27] Pulter clearly recognised the association between emblem poetry and alchemy, but her take on the convention is more naturalistic than allegorical. Emblem 40 begins:

> View but this tulip, rose, or July flower
> And by a finite see an infinite power.
> These flowers into their chaos were retired
> Till human art them raised and reinspired
> With beating, macerating, fermentation,
> Calcining, chemically, with segregation.
> Then, lest the air these secrets should reveal,
> Shut up the ashes under Hermes seal.
> Then with a candle or a gentle fire
> You may reanimate at your desire
> These gallant plants, but if you cool the glass
> To their first principles they'll quickly pass.
> From sulphur, salt, and mercury they came;
> When they dissolve they turn into the same.
> Then seeing a wretched mortal hath the power
> To recreate a Virbius of a flower,
> Why should we fear, though sadly we retire
> Into our cause, our God will reinspire
> Our dormant dust and keep alive the same
> With an all-quick'ning, everlasting flame.
> Then though I into atoms scattered be,
> In indivisibles I'll trust in thee.[28]

---

[27] Quoted from *Emblems and Alchemy*, ed. Alison Adams and Stanton J. Linden (Glasgow, 1998), p. v.

[28] Quoted from *Poems*, ed. Eardley, p. 241; 1–22. See also Eardley's discussion of this poem in 'Hester Pulter's "Indivisibles" and the Challenges of

Pulter is swift to progress from *pictura* to *subscriptio*, but the moral message is fed into a practical guide to alchemical experiment. The emblem-image itself, that of the flower, is indistinct and potentially any one of three flowers we might imagine setting up root in her garden. When Pulter views the flower-as-emblem, she sees beyond the petals themselves to what is beneath and above: a comparison between the alchemical theory of palingenesis – the process of rebirth or regeneration, commonly practised on plants – and the divine 're-inspiration' of mortal dust.[29] She therefore accommodates the spiritual mystery of the 'everlasting flame' within a practice that is both physically knowable to humankind and spiritually resonant. The references to the Paracelsian elements of 'sulphur, salt, and mercury', the classical myth of Virbius, and the 'Hermes seal' demonstrate the voracity of Pulter's reading, yet there is an active, observational quality to her poetry that moves beyond armchair theory.[30] She describes palingenesis as a step-by-step process, reconstructing the physical details of an experiment to create an accessible metaphor for the otherwise ineffable power of divine resurrection. Emblems, according to Bacon in *The Advancement of Learning*, 'reduceth conceits intellectuall to Images sensible'; it is with a similar sense that Pulter informs us to 'by a finite see

---

Annotating Early Modern Women's Poetry', *Studies in English Literature 1500–1900*, 52.1 (2012), 117–41 (121–22).

[29] Thomas Browne describes the process of palingenesis in *Religio Medici*, claiming that it can 'from the ashes of a plant revive the plant, and from its cinders recall it into its stalk and leaves againe'. His thoughts on palingenesis advance to spiritual reflection in a way comparable to Pulter's poem: 'What the Art of man can doe in these inferiour pieces, what blasphemy is it to affirme the finger of God cannot doe in those more perfect and sensible structures?' See Thomas Browne, *Religio Medici* (London, 1643), p. 110.

[30] For a discussion of the Paracelsian elements, see Walter Pagel, *Paracelsus: An Introduction to Philosophical Medicine in the Era of the Renaissance* (Basel, 1982), pp. 100–06; Ovid refers to Virbius the 'minor God' in *Metamorphoses* 15. 545, see David Raeburn, *Metamorphoses: A New Translation* (London, 2004), p. 620; for the Hermes seal ('the hermetic seal which closes the alchemical vessel and keeps it airtight') see Lyndy Abraham, *A Dictionary of Alchemical Imagery* (Cambridge, 1998), p. 99.

an infinite power', as the details of the botanic experiment give evidence for the promise of spiritual rebirth.[31] While she is firm in redirecting the reader's attention to that which can be received through the senses, the aim is not to find all the answers. The 'finite' side of the metaphor is distanced from the opposite that it discloses, thus preserving the mystery of the divine. If 'a wretched mortal' has the power to restore a plant to life, the restorative powers of the immortal deity are comparatively inconceivable. The accommodating image of palingenesis does not solve the mystery but asserts the infallibility of the resurrection on the basis of its ineffability, supporting the conclusion that one must 'trust' in God.

Given the strong impact of the emblematic form on early modern 'chymistry', it is unsurprising that it leads Pulter to her most experimental take on atomic matter. Her 'atoms' provide the central crux for this balancing act between finite and infinite, natural and spiritual, knowable and unfathomable, through which the greatness of God is felt rather than known. Atomism had long played a part in alchemical practice. Pulter's collection of images in Emblem 40 communicate her knowledge, whether at first hand or no, of experimental corpuscular theory from a long tradition of western alchemists, most especially based on the writings of the thirteenth-century scholar known as Geber (hereafter referred to as Pseudo-Geber).[32] It was through alchemy that atomism entered the laboratory; it was alchemy, William R. Newman argues, 'that provided corpuscular theorists with the experimental means to debunk scholastic theories of perfect mixture and to demonstrate the *retrievability* of material ingredients' (italics mine).[33] Pulter appears fascinated with this concept of material 'retrievability', a form of rebirth at the point of utmost obliteration. It is the link

---

[31] Francis Bacon, *The Oxford Francis Bacon IV: The Advancement of Learning*, ed. Michael Kiernan (Oxford, 2000), p. 119.

[32] Geber, or Pseudo-Geber as he is often known, based himself on '"Jabir ibn Hayyan," a semi-fabulous Arabic author who supposedly lived in the eighth century and spawned almost three thousand works'. For more information see Newman, *Atoms and Alchemy*, pp. 26–34 (p. 26).

[33] Newman, p. 3. On the classical atomists as 'strangers to the laboratory', see ibid., p. 25.

between atoms – the 'indivisibles' of the material world – and the indivisibility of divine power that most fittingly supports her argument for human retrievability. Appropriately, given her plural use of the word, that which is 'indivisible' resonates multiple meanings. On the one hand, Pulter's trust 'In indivisibles' refers to her reconstituted physical status following death, finding some comfort in the endurance of her 'scattered' atoms; on the other, importantly, it evokes the superior indivisibility of God's promise of eternal life. The suitability of the palingenesis metaphor for human resurrection is accentuated by the synonymy of terms that collaboratively embody states of generation and corruption: the 'chaos' of the flowers in seed; the 'ashes' and 'first principles' of the same; and the relative 'cause' and 'dust' of the decayed mortal body.[34] Pulter chooses to support her final conclusion not with the timely image of the phoenix rising from the 'ashes', a common emblem within medieval and early modern alchemical works, or with the biblically resonant reference to 'dust' – it is the 'atom' that provides the ultimate object of accommodation for the infinite divine power.[35] Human beings are constructed from atoms which, though dust, are indivisible; what is indivisible can never be destroyed.

In her recent, detailed edition of Pulter's writings, Alice Eardley argues that she referenced atoms only to reject them in favour of 'biblical ideas about dust' and 'an Empedoclean model of the four elements (earth, water, air, and fire)'.[36] The line Eardley draws between the biblically inspired 'dust to dust' and the pagan alternative 'atoms to atoms', however, blurs in the collocations of Pulter's poetry. In a footnote, Eardley cites 'The Revolution', a poem in

---

[34] For more on the alchemical significance of 'chaos' and first principles, see Abraham, pp. 33–34 and 153–56.

[35] The phoenix image was used frequently as a symbol for resurrection and the philosopher's stone in alchemical poetry and prose. For further details, see Abraham, p. 152. Regarding the biblical resonance of 'dust', Genesis 2. 7 refers to the formation of human beings from 'dust' and the circularity of mortal generation and corruption is evoked at 3. 19: 'dust thou art, and unto dust shalt thou return'.

[36] Eardley, in *Poems*, ed. Eardley, p. 26.

which Pulter apparently dismisses atoms in favour of calcined 'dust', the matter inspired by God's breath with the potential to rise again:

> Who can thy infinite power rehearse,
> Which didst create this universe
> And canst to atoms it disperse.
> Should all annihilated be,
> Which is as easy unto thee;
> Oh what would then become of me?
> Nay, rather all to dust calcine;
> I gladly will my form resign,
> It will my carnal heart refine.
> My tears my dust shall rarefy
> To air, which circularly
> Thy blesséd name shall magnify. (p. 110; 19–30)

While it is possible that Pulter could be offering 'dust' as an alternative to a world dispersed into atoms, it seems more likely, especially given her clusters of imagery elsewhere, that the resignation to dust is near-synonymous with the atoms she mentions previously in the poem. 'The Revolution' echoes the vocabulary of corpuscular dissolution from Emblem 40: while the passage above mentions disintegration to atoms and dust, other lines imagine a world falling to 'horrid chaos', turning to 'earth' and calcining 'to ashes' (pp. 109–11, 14, 18, 50). Dust, ashes and atoms accumulate in Pulter's poem to accentuate her focus on physical annihilation, working to set the scene collaboratively rather than separately. It is following her reference to atoms in the above-quoted passage, and her accompanying, fearful question, that the poem marks its abrupt turning point with 'Nay, *rather* all to dust calcine' (italics mine). Pulter's use of 'rather' could be interpreted as a move away from atoms and towards a favouring of dust, but it is more likely an exclamation that emphasises the lyric's sudden change of direction: the poet's realisation that the resignation of form, though at the level of atomisation, paradoxically results in the resurrection and purification of divine being ('It will my carnal heart refine'). This is a familiar pattern in Pulter's poetry, where acts of calcination and atomic dispersal merge in highly personal exercises of devotion. In Emblem 40, the focus on atoms is what prompts the trust 'in indivisibles' (of particles, God and eternal life). Similarly, in 'The Revolution', it is the threat of atomic dissolution that provokes the poet's question about her own fate and reveals the fortunate, surprising answer: that physical destruction is the necessary first stage of the resurrection process. Like Emblem 40, 'The Revolution' mingles alchemical verbs ('calcine'; 'refine'; 'rarefy') with an atomistic setting (dissolution into atoms; resignation of form; reformation) to support the conclusion to trust in the necessity of physical annihilation.

Eardley's observation that Pulter concentrates on 'the Empedoclean model of the four elements' strengthens this reading. This focus can be gleaned from 'The Revolution', where tears absorb dust which they 'rarefy / To air'; a later stanza declares 'higher still I must aspire' and looks ahead to when 'I dilate myself to fire'. The Empedoclean theory contributes significantly to the matter of Pulter's poetry, but it complements rather than replaces her alchemical–mystical take on atomism. In several of his most prominent works including *De Caelo*, *Metaphysics* and *Physics*, Aristotle associates Empedocles closely with the pre-Socratic atomists, a connection that persisted through medieval corpuscular alchemical theories – most particularly in Pseudo-Geber's understanding of the composition of materials not as Aristotelian mixture, but as 'corpuscular juxtaposition'. William Newman explains: 'Empedocles had maintained a century before Aristotle that the four elements were composed at the microlevel of immutable particles, which lay side by side to form compounds (what chemists today would call "mixtures")'; this theory offered a significant contrast to Aristotle's claim that 'genuine *mixis* occurred only when the ingredients of mixture acted upon one another to produce a state of absolute homogeneity'.[37] Pulter is strongly influenced by this Empedoclean–alchemical tradition of corpuscular matter theory, philosophically and, it is tempting to say, aesthetically. Her diverse poetic forms – brief and lengthy; continuous and in stanzas; shifting from pentameter to tetrameter; rhyming in triplets and couplets – reflect the formal diversity of corpuscular juxtaposition in ways that complement their overarching themes: their attempts

---

[37] Newman, p. 28.

to reduce worldly being to its component parts and, from there, to aspire towards a higher elemental status closer to God.

Sadly, no records survive of the contents of the Pulter family library, so any attempts at reconstructing the details of Hester Pulter's reading remain guesswork. Reginald Hine, who compiled a study of the assets belonging to her grandson, James Forester, took greater interest in the cost of the family linen than in their collection of books.[38] It is compellingly possible that Pulter was aware of the Aristotelian corpuscular matter theories of Sir Kenelm Digby, whose English *Two Treatises* was first published in 1645 and was received with great popular interest.[39] Like Pulter, Digby combined the Empedoclean–Aristotelian four elements with atomism; moreover, he placed great emphasis on 'rarity' as a quality in producing natural phenomena. 'Rare' and 'rarefy', as demonstrated in the above quoted lines from 'The Revolution', are two of Pulter's favourite words, appearing time and time again as her poems trace the journey from mortal fragmentation to divine sublimation. In his first treatise, Digby explains that when bodies are rarefied 'the little atomes perpetually move up and downe in every space of the world, making their way through every body, [and] will set on work the little parts to play their game'.[40] Rarity and rarefaction refer to bodies with little density, thin and insubstantial yet incredibly powerful, as Digby's account of atomic infiltration testifies. While the mention of mischievous 'game' resonates more with Cavendish's serious-yet-playful *Poems and Fancies*, the discovery of immense influence and power in dissolution would certainly have appealed to Pulter, for whom the resignation of material 'form' ('The Revolution') results in 're-inspiration' (Emblem 40).

---

[38] Hine's study is the only one that exists of the Pulter family assets. Reginald Hine, *Relics of an Uncommon Attorney* (London, 1951).
[39] Alice Eardley also notes it was likely Pulter was aware of Digby due to parallels in their respective accounts of palingenesis. The Digby text she references on this topic is *A discourse concerning the vegetation of plants* (London, 1661). See 'Hester Pulter's "Indivisibles"', pp. 130–31.
[40] Digby, *Two Treatises*, p. 145.

The purpose of Digby's *Two Treatises* was to construct a mechanical thesis of the body to prove the incorporeality, and hence the immortality, of the soul. His treatise on the body is significantly longer than his section on the soul, which draws its philosophical conclusions deductively from what can be known of corporeal properties. John Henry's summary is accurate:

> The essence of Digby's thesis is to show that the soul is incorporeal and to go from there to demonstrate its immortality. Mortality, involving change, decay, dissolution and so forth, can only apply to the rearrangement in space of the constituent parts of bodies. A separated or 'unbodyed' soul, therefore, must be completely incapable of change.[41]

A Catholic Aristotelian, Digby begins with the certainty of the immortal soul and then seeks to prove it by contrast with the qualities of corporeality. If mortality requires continual change, it follows that an 'unbodyed' constant like the soul must be immortal. Pulter, a devout Anglican, assumes a similar stance. It is her knowledge of the inevitability of corporeal dissolution that supports her trust in spiritual resurrection, a trust anchored in the 'indivisibles' of her mortal frame that permit a glimpse of the eternal life to come. Contra Digby, whose particles were more precisely associated with the tradition of *minima naturalia*, Pulter is adamant in Emblem 40 that the atoms connecting her frail body with the divine are true 'indivisibles'. Her complex chemical, atomistic understanding of matter poses no conflict with her theology. On the contrary, her claim 'though I into atoms scattered be, / In indivisibles I'll trust in thee' goes as far as to strengthen the most important principles of her faith.

Considering this parallel with Digby's concern about what can be gleaned from corporeal structures, it may come as less of a surprise that Pulter, a poet whose ruminations frequently lead her to meditate on the sublimation of body into spirit, does not place much emphasis on the dexterity of fire – the purest, most celestial of the four elements. The heaviest and basest element, earth, is the substance that receives the most attention at her pen. Dust accumulates in her poems: it is clear that Pulter's 'atoms', where they appear,

---

[41] Henry, *Atomism and Eschatology*, p. 224.

are always assumed to be those of corporeal entities – especially those of human bodies. Her poem 'The Invocation of the Elements, the Longest Night in the Year, 1655' offers some explanation as to why earth holds such promise. Pulter composed these lines at a bleak time, a mid-winter night at the darkest point of the Interregnum, a period when there appeared to be little light at the end of the tunnel for subdued Royalist supporters. Most overwhelmingly, Pulter was grieving for the death of seven of her children. She pleads to 'Water': 'Seaven lovly Buds thou hast drawn dry, / Ah, spare the Rest, or else I dye' (p. 140). Beginning with her address to 'Water', Pulter sets up her own hierarchy of elements, with one significant shift away from the Empedoclean order. Earth, rather than first and lowest, is dealt with at the poem's conclusion, implying its position as the element of greatest importance. While the implorations to water, air and fire are written from perspectives of respectful distance and formality – 'Cool christall Water'; 'Sweet Ayr, Refresher of Mankind'; 'Most Noble and Illustrious Fier' – Pulter begins her section on earth with a familiar address, 'Dear Dust' (p. 141), a token of intimacy which echoes across the openings to her poems ('Dear God' and even 'Dear Death'.) Mourning for her children, she concludes the poem with a single plea:

> I ask no pyramid, nor stately tomb;
> Do but involve me in thy spacious womb.
> To beg this once, dear mother, give me leave;
> Oh let thy bowels yearn and me receive. (p. 142; 79–82)

Pulter appeals to 'Dust' as 'dear mother', acknowledging the intrinsic intimacy between mortal being and earth, to whom she writes 'From thee I drew my birth'. There is a tenderness to Pulter's verse that exceeds standard reflection on 'dust to dust' from the Book of Common Prayer. She identifies Earth as female and claims her as mother, seeking dissolution within her 'spacious womb' as a source of comfort and reunion with her lost children. It is not too much of a stretch, based on her natural philosophical imagery in 'The Revolution', to imagine the reconciliation with her children as a corpuscular combination – the promise of reunion through calcination, atomic recongregation and mixture.

Suggestively, the fourth section of 'The Invocation of the Elements' has fourteen lines, an unusual decision for Pulter, who more commonly avoids conventional forms, metrical patterns and rhyme schemes. Another of her poems, likewise appealing to dust and to death, comes tantalisingly close to adopting the sonnet form:

The Hope, January 1665

Dear Death, dissolve these mortal charms
And then I'll throw myself into thy arms.
Then thou mayest use my carcass as thou lust
Until my bones (and little luz) be dust.
Nay, when that handful is blown all about,
Yet still the vital salt will be found out,
And when the vapor is breathed out in thunder,
Unto poor mortals' loss, or pain, or wonder,
And all that is in thee to atoms turned,
And even those atoms in this orb [are] burned,
Yet still, that God that can annihilate
This all and it of nothing recreate,
Even he, that hath supported me till now,
To whom my soul doth pray and humbly bow,
Will raise me unto life. I know not how. (pp. 180–81)

This poem again opens with an intimate appeal, a prayer not to God, as one might initially expect, or even to 'Dust' – it is to 'Death' that Pulter vows 'I'll throw myself into thy arms'. What was tender and maternal in her relationship with Dust is here dark and troubling, with a glimpse of ecstatic relief in the surrender to annihilation and a disturbing layer of violence ('use my carcass as thou lust'). Chemical experiment again enters the remit of Pulter's poetry to enrich the declaration of her faith, as 'The Hope' reimagines the created cosmos as alchemical vessel, an 'orb' in which the Paracelsian 'vital salt' of bodies is purified through 'vapor' and fire. Doubled with the strong chemical language of the poem, clearly supported by some nuanced knowledge of alchemical philosophy, is the promise of Doomsday: thunder, fire and utter devastation, to the extent that even the 'little luz' – an indestructible bone in the body, thought to be based at the top of the spinal cord – would dissolve to dust. According to Judaic legend, the luz was the source of the body's resurrection at the final reckoning.[42] Pulter's desire for total dissolution is so strong that she explicitly destroys that which is supposedly indestructible.

This desire for annihilation extends, however, to Death itself. When Pulter writes, addressing Death, 'all that is in thee to atoms turned', her sense carries more than one possible meaning. She may be indicating, in accordance with the progression of her poem from personal to universal dissolution, that Death ultimately absorbs all things and reduces them to atoms. The pointed specificity of 'in thee', however, suggests the destruction of her addressee. Pulter hints at the defeat of Death as she approaches the volta of her poem, which moves away from her initial correspondent and towards God, in whose promise of resurrection she trusts. As in Emblem 40 and 'The Revolution', the reference to atoms is what triggers the turning point and reminder of spiritual 'Hope'. One key difference is that 'The Hope' places no limits on its imagined physical dissolution (unlike the other poems, which stop at atoms): it is not just the created world, but its very atoms that are 'burned'. The annihilation of these atoms directs Pulter towards her 'Hope', as the wondrous ability of God to 'recreate' even from nothing – *ex nihilo* – is held as evidence of the divine power that will enable her resurrection, though she knows 'not how'. In this final line, the mood of trust resonates with her claim from Emblem 40, '[i]n indivisibles I'll trust in thee' – a clause flooded with negative 'in' prefixes that shares an emphasis on negative theology, of ultimately knowing through not-knowing.[43] The final triplet of 'The Hope', as Alice Eardley has

---

[42] For further details see *Poems*, ed. Eardley, p. 180. See also the note on 'The "Little Luz"' by Elizabeth Scott-Baumann, from *The Pulter Project*: <http://pulterproject.northwestern.edu/poems/ee/the-hope/#the-little-luz> [accessed 26 Feb. 2021].

[43] Alice Eardley comments further on Pulter's closing 'I know not how': 'This is not the first place Pulter introduces a dramatic hiatus just at the moment when a particular reverie reaches a point at which it cannot, or should not, go, usually, as in this instance, when it begins to encroach on divine mysteries.' See Eardley, '"I haue not time to point yr booke … which I desire you yourselfe to doe": Editing the Form of Early Modern Manuscript Verse', in *The Work of Form: Poetics and Materiality in Early Modern*

observed, breaks 'beyond the limits of the couplets regulating the poem as a whole'.[44] By resisting the fourteen-line sonnet structure and Pulter's characteristic couplets, the closing lines of the poem literally transform in a way sympathetic to the future, imagined re-formation of the resurrected poet. Her reduction to atoms, and the burning of these in turn, initiates the alchemical purification of being that makes eternal life possible.[45]

A prominent message emerges from Pulter's 'atom poems'. She discovers divine trust in the promise of world-breaking rather than world-making, paradoxically seeking comfort in the idea of rebirth through thoughts of terrestrial obliteration. Her most positive conclusions are dealt through negation. The final line of 'The Hope' is exemplary English heroic meter in the stead of Pulter's frequent irregularities, confident and assertive in its iambic pentameter and well-placed caesura. Consequently, and complicatedly, the closing 'I know not how' raises different possibilities: it is potentially sceptical of the possibility of resurrection, following the annihilation even of indivisible things (atoms and luz); it is unsure, hopeful and trusting in divine power; and it is a strong monosyllabic statement that stresses the conviction and certainty of her faith. Divine trust is the ultimate message, but it takes meditation on the extremities of annihilation to get there. Pulter discovers in the atom, the base

---

Culture, ed. Ben Burton and Elizabeth Scott-Baumann (Oxford, 2014), pp. 162–78 (p. 166). Scott-Baumann also quotes from Eardley's essay in her note on 'Knowledge, Faith and Doubt' for 'The Hope' in *The Pulter Project*, <http://pulterproject.northwestern.edu/poems/ee/the-hope/#knowledge-faith-and-doubt> [accessed 26 Feb. 2021].

[44] Eardley, 'Editing the Form of Early Modern Manuscript Verse', p. 166.

[45] The message is accentuated by comparing 'The Hope' to another lyric Pulter addresses to Death, 'The Welcome [2]', in which the speaker lays down:

> In dust as in the daintiest bed of down
> Where I to my first principles must turn
> And take a nap in black Oblivion's urn
> Until the sun of life arise in glory
> And then begins my everlasting story. (p. 129, 6–10)

Pulter's reference to 'first principles' is another example of her conflated atomic and alchemical imagery.

material of all things, a powerful model for the cycle of dissolution and resurrection. Through the scattering and even, in 'The Hope', obliteration of atoms, she draws attention to the promise of divine indivisibility in what comes next. Physical corruption is the most positive (while remaining fathomable) stage of the journey to rebirth.

While Emblem 40 affirms this argument through an observational account of natural philosophical experiment, Pulter's 'Poems' take a more personal, subjective interest in the spiritual consequences of atomic being. Poem 58, quoted at the beginning of this chapter, succinctly contracts her trust in negation and the transient imperfection of embodiment:

> For I no liberty expect to see
> Until to atoms I dispersèd be.
> Then being enfranchised, free as my verse,
> I shall surround this spacious universe,
> Until by other atoms thrust and hurled
> We give a being to another world. (p. 169)

'For I no liberty' begins, like many of the pieces in Pulter's collection, *in medias res*.[46] Her verse dives headlong into the conclusion of its argument, her voice drawing a strong line under contentions of life and death in a way that becomes typical of her poetry. Here, as in many of her poems, the content moves swiftly away from what should be, at first glance, its central message: the focus is less

---

[46] Pulter's poems often begin as though in the middle of a conversation. Examples of first lines creating this impression include 'On Those Two Unparalleled Friends, Sir G.[eorge] Lisle and Sir C.[harles] Lucas' ('Is Lisle and Lucas slain? Oh say not so'); 'My Love is Fair' ('And is thy love so wondrous fair?'); and Poem 64 ('And must the sword this controversy decide'). See *Poems*, ed. Eardley, pp. 73, 170 and 179. The editors of *The Pulter Project* consider 'For I no liberty' not as a separate poem, but as a 'six-line coda to "Why Must I Thus Forever Be Confined" (Poem 57)'. See Victoria A. Burke, 'Speculations about Multiple Worlds', <http://pulter-project.northwestern.edu/poems/ee/why-must-i-thus-forever-be-confined/#speculations-about-multiple-worlds> [accessed 26 Feb. 2021]. From this reading, 'For I no liberty' presents the conclusion to the opening question of Poem 57: 'Why must I thus forever be confined / Against the noble freedom of my mind?' (p. 164, 1–2).

on the conviction of what will put an end to imprisonment, and more on what will occur after Pulter sheds off her mortal coil. The word 'Then' is one of her regular signposts. She directs her reader to the spiritual, theological reading of an initially physical subject; moreover, she opens – 'enfranchises' – the pithy space of her verse and expands to encapsulate divine significance. This is enabled by her empathy with atomic being, an intimacy that is itself made possible by the poetic setting: her reasoning smoothly transitions from rational logic to an intuitive trust in divine mysteries that is empowered by her 'free ... verse'. Atoms and poems set Pulter free.

Her allusions to atoms often result in interrupted cadences.[47] What initially seem to be complete resolutions are hastily followed by interjections which, by introducing additional material or further conjectures, draw her work out towards its final, Christian conclusions. Declarations of how worlds and beings 'disperse' into atoms are generally followed by scene-changing adverbs – 'then' or 'yet' gesture at how atoms expand the focus of Pulter's poems, permitting her explorations into the deeper mysteries of the faith. The ecstatic dissemination of particles is resonant of Kenelm Digby's 'little atomes' that 'perpetually move up and downe in every space of the world': in 'For I no liberty', the apparently ultimate stage of dissolution – the dispersal of atoms – becomes the ultimate liberation of self. Pulter scatters her particles to the far corners of the universe, and in doing so paradoxically discovers her 'being' while loosening her former identity amongst the surrounding atoms of the cosmos. There is great confidence in this poem; a higher level of assertion than in many other of Pulter's devotions, which rest on humility and hope. With 'I dispersèd *be*', true being occurs following the dissolution of the 'I' and the confidence of 'We give a being to another world' asserts a collective power. The exact nature of this 'we' remains ambiguous. Congregating into new formations of matter, the atoms of Pulter's new world on the one hand belong to the resurrected dead, but on the other emerge from a recombination of elements from inert, gross material entities as well as

---

[47] I am indebted to Emma Pauncefort for this observation, 9 May 2016.

those which had previously constituted sensitive and rational life.[48] Yet there is also the possibility that Pulter's 'we' might not be inclusive of any external individuals at all. Later in her manuscript, she concludes Emblem 44:

> Nor shall my scatter'd dust forgotten rest,
> But like the embryo in the phoenix nest
> That word that nothing did create in vain
> Shall reinspire my dormant dust again
> And from obscurity my atoms raise
> To sing in joy his everlasting praise
> And reunite my body to my spirit
> That we may those eternal joys inherit,
> Which I may claim by my dear savior's merit. (p. 250; 30–38)

What begins in the first person singular enters into the plural in the penultimate line. The nature of the 'we' is uncertain but may not incorporate other human beings. An ambiguity springs between a 'we' that might refer to Pulter's own multitude of atoms, 'other atoms' from external objects, or the dual components of her sensitive and rational composition – body and soul. In light of this, 'For I no liberty' might be looking ahead to the reconciliation of these two aspects of the human condition.

At this point, strains of mortalism could seem to interfere with Pulter's displays of Christian humility. Emblem 44's reinspiration of 'dormant dust' hints not gently at mortalist views – of the moderate (nevertheless deemed radical and heretical) belief in 'soul-sleeping'; the death of the body and soul until resurrection at the Last Judgement.[49] Whether Pulter herself adhered to mortalism is unclear,

---

[48] Louisa Hall offers an insightful analysis of this plural act in 'Hester Pulter's Brave New Worlds', pp. 181–82. She argues: '"Infranchis'd," Pulter's atoms are simultaneously disaggregate and aggregate, particulate and wavelike, unmistakably her and also part of a larger unity' (p. 182).

[49] For more information on Christian mortalism during the seventeenth century, see Bryan Ball, *The Soul Sleepers: Christian Mortalism from Wycliffe to Priestley* (Cambridge, 2008); Norman T. Burns, *Christian Mortalism from Tyndale to Milton* (Harvard, 1972); Richard Sugg, *The Smoke of the Soul: Medicine, Physiology and Religion* (Basingstoke, 2013).

and rather unlikely: the principle comes into conflict with other expressions of her belief and would have little place in an otherwise orthodox Anglican faith.[50] The strains of the phenomenon in her work are symptomatic of her dependence on the fragmentary body as the location of meaning, the medium by which she comes to 'claim' her eternal joys in God. Richard Sugg has written on the power of mortalism as 'literary fantasy' in seventeenth-century England, arguing of Donne:

> His mortalism is not a matter of overt and settled theology, but of personal fear and personal longing... [M]ortalism was attractive because of the way it offered a solution to the temporary loss of the body – an entity on which Donne's self seems to have been peculiarly dependent.[51]

The mortalist themes in Pulter's writing appear to have similar force. Perhaps the corporeal requirement Sugg observed in Donne is not so 'peculiar' after all: Pulter shared with Donne a dependence on reading the body – especially the failing, crumbling body, accompanying an 'afflicted soul' – as the medium through which divine truths could be made known.[52] She moreover sought meaning and spiritual value through the body reduced to its atomic parts. The fate of 'dissolution', one of Pulter's most-used words, is a spiritual sign from which she deduces the omniscient power of the divine.[53] The dissolved and scattered body also invokes a multiplicity of being – a prevalent identity beyond the single embodied self – that strongly appealed to Pulter's conceptualisation of the

---

[50] Pulter's Anglicanism comes to the fore in her Laudian moments and belief in a Church headed by the monarchy. Pulter attacks Presbyterians and Independants in Emblem 52: 'So this sad kingdom locusts did o'errun; / Such clouds (ay me) as did eclipse our sun. / What house of base vermin then were free? / Such a like army let me never see' (pp. 260–61, 14–17).

[51] Sugg, pp. 234–35.

[52] For 'afflicted soul', see 'The Invocation of the Elements', line 1 – but such allusions to the soul are commonplace across Pulter's work. When evoked, her soul is 'dark', 'afflicted', 'struggling'.

[53] Pulter refers to 'dissolution' throughout her poems. For just a few examples (there are more), see 'The Eclipse', 'Aurora 1', 'Universal Dissolution', Poem 40 and Emblem 40 (Eardley, pp. 46, 56, 65, 137 and 231).

ideal Christian subject. Many of her verses, including both Emblem 44 and 'For I no liberty', follow the act of corpuscular dispersal and conclude by contracting multiplicity in singularity. The 'we' of her emblem joins with the 'I' of the final line, and 'eternal joys' become an obtainable possession thanks to the 'savior's merit'. In 'For I no liberty', the plural combines to produce a unified, comprehensive creation, the 'being' of 'another world'.

This relationship between atomic being and the shifting status of pronouns taps into one of Pulter's key interests in the atom. From imagining the disintegration and reformation (in both senses) of her corporeal body, she moves to associate her shifting perceptions of self-identity with the movement of indivisible particles. Her personal pronouns change form with her atoms, resulting in a selfhood that is insecure, but positively flexible and free. She trusts in the dissolution and disfigurement of mortal form because it anticipates the promise of divine repair, a spiritual rebirth figured in atomic and alchemical terms. This is the dominant theme of Emblem 44. Pulter bases her emblem on the ancient story of the 'Brahman', Calanus, teacher of Alexander the Great. According to Plutarch's account, upon travelling west with the Macedon king, Calanus fell chronically ill with dysentery and ended his own life: on a 'funeral, flagrant pile he lies; / Becoming thus both priest and sacrifice' (Pulter, p. 249; 5–6).[54] Pulter observes in Calanus's immolation the regenerative self-sacrifice of the phoenix in her nest, who dies in order to be reborn as 'embryo' ('like the embryo in the phoenix nest / That word that nothing did create in vain / Shall reinspire my dormant dust again') – a curiously loaded word for her to fall back on, as that which was 'embryon' in seventeenth-century English could signify not only unborn potential but imperfect formation.[55] Thomas Blount defined '*Embryon (embryo)*' in his *Glossographia*

---

[54] For Plutarch's account of the life of Calanus, see *Vita Alexander*, in *Plutarch: Lives*, trans. Bernadotte Perrin (London, 1959), p. 69. While the story of the death of Calanus was familiar to many, its representation as visual emblem was not.

[55] As recalled by the ambiguous significance of Milton's troubled 'embryon' atoms in *Paradise Lost*, 2. 898–900. See Chapter Four of this book, p. 175.

of 1661 as 'a child in the mothers womb, before it has perfect shape; and by Metaphor, any thing before it has perfection'.[56] The second part of Blount's definition has especial resonance with Pulter's poetry, which discovers hope and divine promise in the necessary imperfections – and shifting atomic constituents – of human and poetic form. For Pulter, nothing has 'perfect shape' other than that which has been 'reinspired' by God, but that which is not fully formed, or which is humble in its weakness and imperfection, has the blessing of pre-accommodating the promise of its future perfection by the will of God.

For Pulter, there is a clear parallel between 'embryon' and atomised matter. Unfixed and imperfect mortal forms are blessed simply for their *need* of divine intervention. If the mortal human form were permanent and at the limits of its perfection, there would be no opportunity for its spiritual growth toward the obtainment of 'eternal joys' (Emblem 44). Pulter's understanding of embryonic being has biblical connotations. Psalm 139 wonders at the omnipresence of the divine, and exclaims (in the words of the King James Bible, quoted from Henry Ainsworth's commentary on the Psalms from 1644):

> [13] For thou hast possessed my reines, hast couered me in my mothers wombe.
> [14] I will confess thee, for that fearefully, maruellously made am I; maruellous are thy works, & my soule knoweth it very well
> [15] My Bone was not hid from thee, when I was made in a secret place, was embrodered in the nether places of the earth
> [16] Mine vnformed substance thine eies did see, & in thy booke all of them were written, in the dayes they were formed, & when not one of them was.[57]

Ainsworth offers a particularly close reading of verse 16 in his commentary. He glosses 'vnformed substance' with the alternative description of '*Mine embryon,* which is *the body in the wombe before*

---

[56] Thomas Blount, *Glossographia* (London, 1661), sig. P2v.
[57] Henry Ainsworth, *The booke of Psalmes, Englished both in prose and metre with annotations, opening the words and sentences, by conference with other Scriptures* (Amsterdam, 1644), p. 160.

it hath perfect shape, or, *unwrought up,* as the Greeke here translateth it. The Hebrew name is of wrapping or winding up like a bottome, *my wound-up masse,* or *body.*' The 'embryon' here accommodates both the 'body in the wombe' and the concept of 'unwrought' personhood. A 'bottome' is a core on which to wind thread; human beings are gradually spun to corporeal perfection, but their creator foresees the potential of their embodiment long before its completion. The slightly enigmatic 'in thy booke all of them were written' is explained by Ainsworth as 'all my members wound up in that my embrion or unperfected substance. Or generally, *all men.*' In the first gloss, the human figure takes form as though it were wound over a pre-existing structure that prescribes its shape; according to the second, in a very Pulter-like sense, *un*winding comes across as the productive creative action. This 'embrion', or 'unperfected substance', already contains – in potential – everything that will come to pass for that individual.[58] Moreover, the metaphor extends to cover not only a single body but '*all men*' represented by the same substance.

The obscurity of the plural in verse 16 of Psalm 139 is in nature similar to the purposeful, suggestive ambiguity of Pulter's shifts in personal pronoun. '[T]hem' can refer to the parts of the individual body of the psalmist, but this does not contradict the possibility of its significance as '*all men*'. It is no coincidence that Pulter returns to the concept of embryonic being frequently in her poems. The promise held by this as-yet-incomplete form is especially prominent in poem 39, headed by the suggestive title 'The Perfection of Patience and Knowledge'. Pulter begins:

---

[58] Strikingly, this biblical claim finds sympathy in Pierre Gassendi's atomic principles of embryonic form. Catherine Wilson writes of Gassendi's 'conviction that human and animal life, generation, and mentality could be described and analyzed in common terms' ('I give a soul to semen; I restore reason to animals; I find no distinction between the reason and the imagination'), *Epicureanism at the Origins of Modernity* (Oxford, 2008), p. 26. This also chimes with Cavendish's sentient philosophy of matter, where the distinctions between rational and imaginative thought break down due to their mutual dependence on the atoms that give them shape.

> My soul, in struggling thou dost ill;
> The chicken in the shell lies still,
> So doth the embryo in the womb,
> So doth the corpse within the tomb,
> So doth the flower sleep in its cause,
> Obedient all to Nature's laws. (p. 135; 1–6)

In a voice seemingly a far cry away from the assertive first-person of 'For I no liberty', she convinces her soul not to strain against its 'bonds' (p. 135, 8). The transition in thought between the embryo, corpse and dormant flower, all likewise 'still' and 'Obedient', patterns out a typical cluster of imagery in the context of Pulter's poetry. Womb, tomb and 'cause' – seed – appeal to her imagination as containers, or points of contraction, for spiritual potential. These are further endorsed by the 'bonds' of her couplets, which tie images of as-yet-unperfected states of being into perfectly measured poetic form, a tension that accentuates her faith in the desirability of fragmentation. For Pulter, atoms – or 'Dust' from the maternal earth with which she feels such close affinity – are the ultimate embryonic matter as they encapsulate the potential of all that is and all that shall be. 'The Perfection of Patience and Knowledge' does not evoke atoms by name, but continues in a manner parallel to the promise of poem 58:

> The chirping bird will break its shell,
> The infant leave its loathéd cell,
> The sleeping dust will rise and speak,
> And will her marble prison break,
> The flower her beauty will display,
> And my enfranchised soul away
> Beyond the sky will take her flight
> And rest above the spheres of night (pp. 135–36; 11–18)

Pulter draws again on one of her favourite words, 'enfranchised', to bring a degree of social and political depth to her declaration of personal liberty. The 'sleeping dust' that 'will rise and speak' recalls the oft-quoted living dust of 1 Corinthians 15.51–52, which promises to rise in the 'blink of an eye', at the 'sound of the trump'; Eardley follows this by glossing all Pulter's references to dust with the biblical sense of 'material from which the human body is derived and to which it will ultimately return'.[59] There is, however, an additional vitality to Pulter's 'dust'. Her sentient dust particles are like her atoms, which are not restricted to or by 'this dunghill globe of earth' (Pulter's derogatory label for the terrestrial world in poem 39). Dust/atoms are not disposed at the act of resurrection, but accompany the 'enfranchised soul' into eternal life and the creation of new worlds.

Pulter encourages and meditates on the dissolution of the current world in order to find her liberty in the next. The world-making of her atom poems rests on acts of world-breaking. It is at the stage of utter dissolution that her indivisible atoms show their true, divine potential; all living matter is, for Pulter, in its 'embryon' state, and comes into its maturity when it is reformed (re-formed) anew. The entire principle of her faith is compressed into Poem 58, a declaration of belief that binds the fate of the soul closely to that of the body, and employs the atom as a quasi-material, quasi-spiritual medium between the two. Pulter knows 'not how', but knowledge is not necessary when she can trust 'in Indivisibles'. As the repeated sound of the negative prefix in this phrase suggestively implies, it is in loss and lack that Pulter finds her answers. Her intimate acquaintance with nothingness directs her to the far reaches of the universe, the point at which she can reach out in hope toward the promised eternity on the other side.

MARGARET CAVENDISH: ATOMS AND FANCIES

> Small atoms of themselves a world may make,
> For, being subtle, every shape they take.
> And as they dance about, they places find;
> Such forms as best agree make every kind.
> For when we build a house of brick or stone,
> We lay them even, every one by one:

---

[59] For an analysis of the significant relationship between this passage and the atom, see Chapter Four, pp. 193–96.

And when we find a gap that's big or small,
We seek out stones to fit that place withal.
For when as they too big or little be,
They fall away and cannot stay, we see.
So atoms as they dance find places fit;
They there remain, lie close, and fast will stick.
Those which not fit, the rest that rove about
Do never leave, until they thrust them out.
Thus by their several motions, and their forms,
As several workmen serve each other's turns.
And so by chance may a new world create,
Or else, predestinate, may work by Fate.

('A World Made by Atoms')

Cavendish's poem appears to pick up where Pulter left off in 'For I no liberty'. Initially, her focus on atomic world-making seems less personal than Pulter's, dwelling not on the hope of future regeneration but on present-tense observation of atomic behaviour. Her verses verbalise a form of philosophical experiment – the first words mark out the stages of her poeticised hypothesis line by line ('And… For… So… Thus'). Nevertheless, the tone of the poem distinguishes her task from straightforward scientific report. Cavendish is aware of the ever-growing dominance of vision as tool in natural philosophical experiment, but is highly critical elsewhere – later in *Poems and Fancies* as well as in her *Observations upon Experimental Philosophy* – of those who rely on the senses to give evidence for discoveries.[60] Instead, the visual analogies of

---

[60] In *Observations upon Experimental Philosophy*, Cavendish argues that 'sense, which is more apt to be deluded by reason, cannot be the ground of reason'. She concludes, therefore, 'that experimental and mechanic philosophy cannot be above the speculative part, by reason most experiments have their rise from the speculative'. *Observations*, p. 49. See also *Poems and Fancies*, 'The Objects of Every Sense Are According to their Motions in the Brain' (Part I), which explores the eccentricity of the senses and concludes 'Imaginations just like motions make, / That every sense is struck with a mistake.' For more information on visualisation in early modern experimental science, see Stuart Clark, *Vanities of the Eye: Vision in Early Modern European Culture* (Oxford, 2007), pp. 329–56.

'A World Made by Atoms' are creative attempts to explain things unseen. The example of building 'a house of brick or stone' incorporates the reader into the natural philosophical and poetic exercise, demonstrating the otherwise unfathomable theory that atoms 'of themselves' might create all things. The poem remains in the mood of possibility – atoms 'may' make a world; they 'may a new world create'. Rather than assert a proposition, Cavendish plays with an idea, its premises and its consequences, and works out through a kind of imaginative reasoning where it might take her.

'A World Made by Atoms' is the first of her fifty 'atom poems' and hints at what lies in store in the ensuing collection. The poet acknowledges that atoms open the 'chance' of a 'new world': the shifting constituents of matter present infinite possibilities, as the dancing particles playfully – but not unseriously – maintain the prerogative to form things anew. But the 'chance' to which Cavendish refers may be no random, senseless possibility. It is just as possible that these atoms are 'predestined' to work by 'Fate'. With this remarkable conclusion to the opening 'atom' poem, Cavendish plucks out a verb with notable Calvinistic connotations. It is not as though atomism and predestination had never come into contact. Pierre Gassendi and Walter Charleton had attempted to Christianise atomic theory by arguing for God's control over the particles of creation; even Lucy Hutchinson, staunchly puritanical and heavily contemptuous of that 'dog' Lucretius, tapped into the sympathies between predestination and the recongregation of atomic matter.[61] Cavendish, however, employs no caution in playing with this association to imply that, in tandem with the creator of atoms at large, the one actively 'predestining' in this case is her, the author of these 'new worlds' in poetry. This sense is heightened by a textual variant in the first edition of *Poems and Fancies*, where 'by Fate' reads as '*my* Fate'. As Liza Blake has observed, this was likely to have been an error: Cavendish's 1664 and 1668 editions both read 'by Fate', and two known copies from 1653 contain handwritten notes that change the 'my' to 'by'.[62] Nevertheless, even without the

---

[61] For further details, see Chapter Four.
[62] See note 13 on 'A World Made by Atoms', in *Margaret Cavendish's Poems and Fancies*, ed. Liza Blake. See textual notes at: <http://library2.utm.

possessive pronoun in the later editions, the atoms of the world are also the atoms, or elements, of Cavendish's text, the combinations over which she asserts authorship. From the very start of her 'atom poems', Cavendish seeks to take charge but, crucially, the power of her verses is in their very openness to possibility and their resistance to fixing natural philosophical meaning. Her atoms have a life of their own: they 'dance about', form productive and destructive relationships, and it is left ambiguous as to whether they operate by 'chance' or by predestination. Here and elsewhere across *Poems and Fancies*, the mood of creative, imaginative possibility emerges as the most suitable medium for natural philosophical enquiry.[63]

Atomism captured Cavendish's imagination, and this imagination in turn inspired her developing philosophies of matter. To understand better the creative appeal of the 'atom poem' it is worth reviewing her attitude to the place of 'Fancy' in writing. She gives the title 'Fancies' officially to Part III of her first publication, but

---

utoronto.ca/poemsandfancies/2017/06/09/a-world-made-by-atoms/> [accessed 26 Feb. 2021]. Blake acknowledges the handwritten changes in copies of *Poems, and fancies* (1653) housed at the Beinecke Rare Book and Manuscript Library and the Bibliothèque Nationale de France. She also notes, in her introduction to the collection, that Cavendish rarely uses the first-person pronoun in her atom poems: 'despite the huge amount of persona-building that happens in the prefatory materials, the poems of *Poems and Fancies* themselves are remarkable for their almost studied avoidance of the first-person singular pronoun.' See 'Reading Poems (and Fancies): An Introduction to Margaret Cavendish's *Poems and Fancies*', <http://library2.utm.utoronto.ca/poemsandfancies/introduction-to-cavendishs-poems-and-fancies/> [accessed 26 Feb. 2021].

[63] See also Jessie Hock, who sheds further light on Cavendish's sceptical, imaginative method by comparing *Poems and Fancies* with Lucretius's *De Rerum Natura*: 'Certainty could be achieved by closer observation, which is unfortunately impossible; with no access to a more privileged perspective on the true causes of things, humans have to make do with "probabillitie." Multiple explanations, then, offer a menu of probable causes from which the reader can choose.' Hock proceeds to describe *Poems and Fancies*, rather beautifully, as a 'choose-your-own-adventure story'. Jessie Hock, 'Fanciful Poetics and Skeptical Epistemology in Margaret Cavendish's *Poems and Fancies*', *Studies in Philology*, 115.4 (2018), 766–802 (778 and 780).

the act of fancy itself shapes all her poems and was instrumental to developing her philosophy.[64] For Cavendish, 'Fancy' is an original act of inspiration; it is generative rather than imitative of some pre-existing idea. It is the most important word in her vocabulary for establishing her aesthetic, though any attempt at definition exposes its inconsistent use across her works.[65] She expands upon the importance of 'Fancy' in the opening message 'To the Reader' from *Natures Pictures* (1656), a collection of short poems and moral tales in prose:

> for descriptions are to imitate, and fancy to create, for fancy is not an imitation of nature, but a naturall Creation, which I take to be the true Poetry: so that there is as much difference between fancy, and imitation, as between a Creature, and a Creator.[66]

The appeal to the reader sees Cavendish confusing her terms, as fancy is both 'naturall Creation' and 'Creator'. It occupies a shifting position in her fused philosophies of matter and of literary authorship. Fancy is the 'Creator' of innate, independent thoughts, but also the medium for presenting thoughts for the perception of others, in which sense it is a 'Creation'. As a 'naturall' creation, an original in nature, the faculty of fancy is also a material substance both answerable to and influential over the processes of natural philosophy. It is intimate with the creative powers of nature and beyond the limited focus of scientific empiricism.

The act of creating new worlds from atoms – whether poems or particles – both mirrors and actualises Cavendish's authorial potential. Despite this, she frequently dismisses her philosophical poems as insubstantial, or frivolous. *Poems and Fancies* opens with

---

[64] She writes in her preface 'To the Reader' at the beginning of the section: 'I desire all those which read this part of my book to consider that it is thick of fancies, and therefore requires the more study.' Blake, <http://library2.utm.utoronto.ca/poemsandfancies/category/section/part3/> [accessed 26 Feb. 2021].

[65] When used disparagingly it is near synonymous with the irresponsible side of 'Fiction' masquerading as 'Truth'; in its more general and positive sense, it refers to the original form of an innately produced idea.

[66] *Natures Pictures*, sig. C3b. All further quotations from this work will be taken from this edition and cited in the text by page number.

the following defence 'To Natural Philosophers': 'the reason why I write it in verse is because I thought errors might better pass there than in prose – since poets write most fiction, and fiction is not given for truth, but pastime'. Though the association she creates between 'fiction' and 'pastime' is designed to demote her poetical efforts to the status of inconsequential – and harmless – hobby, her admittance of permitted 'errors' in her fiction has a double effect. If poetical fiction had nothing to do with the expression of 'truth' or expounding of philosophical theories, there would be no possibility of 'errors' within it. The poems of *Poems and Fancies* stress the hypothetical nature of their claims in the cautiousness of their tone: repetitions of 'if', 'may', and uses of comparative simile allow for errors if they happen to occur, but also convey a more sensitive proposal of a truth that Cavendish is unable to test beyond verbal expression. The flipside to this is that verbal expression comes closest to finding the answers. If there were no possibility of communicating 'truth' in fiction, she would not have revisited her *Poems and Fancies* ten years after its initial publication to amend its explanatory annotations and the organisation of its philosophical content.[67]

Cavendish's defence that her poems are insignificant in relation to her natural philosophy is, therefore, unconvincing. Words like 'fiction', 'fancy' and 'wit' remain vague and often interchangeable throughout her writings, but this very flexibility of vocabulary – the connotations of words which expand and diminish according to the context in which they appear – is a defining aspect of her enquiring but sceptical philosophy. While the author explicitly lays out the negative consequences of trusting a 'fictional' theory as fact, she attributes positive and philosophically productive purposes to the imaginative faculties. Cavendish's method of philosophising embraces ambiguity and possibility and, as revealed by the hypothetical propositions of her poems, these characteristics are well-suited to the creative bounds of imaginative writing. Her dual approach to fictional reasoning and reasoning in fiction appears more clearly in her criticisms of the experimental science practiced by the Royal Society. The fundamental incompatibility between her speculative thought and the empirical methods of Robert Hooke

---

[67] I discuss these alterations in a later section of this chapter, pp. 166–69.

and Henry Power centres not upon the trying out and testing of theoretical ideas, but on her criticism of the scientists' complacency in drawing 'discoveries' from their experiments.[68] Telescopes and microscopes in the philosopher's laboratory mislead the experimental scientist into trusting the images they present as the 'real body of the object'. Cavendish protests that Hooke's infamous blown-up picture of a louse in *Micrographia* (1665) is nothing more than a monstrous imitation of nature: 'the glass only figures or patterns out the picture presented in and by the glass, and there mistakes may easily be committed in taking copies from copies' (*Observations*, p. 52). She continues to describe microscopy as an 'unprofitable art', capable only of false results. It 'at best produces mixt or hermaphroditical figures, that is, a third figure between nature and art' (p. 53).

The 'hermaphroditical' figure of the louse, a disproportionate image not only contrary to nature but also alien to the characteristics of 'art', represents for Cavendish the negative side of fiction, what she refers to disdainfully in her *Philosophical Letters* as 'meer Poetical Fictions, and Romancical expressions, making material Bodies immaterial Spirits'.[69] By relying on such an image as empirical proof, she argues, the scientist has fallen on sterile ground. Despite her criticisms, however, Cavendish practises her own method of 'experimental philosophy' that has everything to do with 'Fiction'. According to her vitalist view of natural construction, every object – inanimate or animate, animal, vegetable or mineral – owns individual qualities of perception that no other perceiving entity can aspire to know.[70] Ambiguous possibilities are

---

[68] Cavendish reacts against Robert Hooke's *Micrographia* (London, 1665) and the Royal Society experimenter Henry Power's *Experimental Philosophy* (London, 1664), another work to focus on the practice of microscopy. For more on Cavendish's reaction against microscopy and experimental science, see Elizabeth Spiller, *Science, Reading, and Renaissance Literature: The Art of Making Knowledge, 1580–1670* (Cambridge, 2004), pp. 137–77.

[69] *Philosophical Letters*, p. 12.

[70] For critical work on Cavendish's vitalism, see John Rogers, *The Matter of Revolution: Science, Poetry and Politics in the Age of Milton* (Ithaca, 1998), pp. 177–211, Jay Stevenson, 'The Mechanist-Vitalist Soul of Margaret Cavendish', *Studies in English Literature 1500–1900*, 36 (1996), 527–43, and

therefore the closest to truth the philosopher can ascertain. In the *Observations* she attributes agency to the 'glass' because it is the lens, and not the philosopher peering through and observing the form, that patterns out the original image of the louse. The philosopher can only perceive at second hand what is perceived by the glass. Cavendish dismisses the practice as 'unprofitable' because it would be impossible to produce further observations from the false copy. Hooke presents the illustration as empirical fact when it is merely a second-hand derivation, but neither is it a work of art since art, for Cavendish, must always be generative, not imitative.

According to Cavendish, the faculty of 'fancy' is a prerequisite of philosophical maturity – simultaneously a means for furthering the understanding and preserving a healthy scepticism in the positing of hypotheses. Sensitivity to the often uncertain distinctions between true and fictitious theories makes for more productive philosophy. This is the juncture at which her poems and short stories make contact with scientific reasoning. Though Cavendish drops atomism as her central subject after Part I of *Poems and Fancies*, her atom poems were at the heart of her developing authorial voice and theories of matter.[71] In a recent article, Jessie Hock has argued that Cavendish draws on Lucretian ideas in *Poems and Fancies* to conceive of 'poetic fancy – which she understands as the imaginative expression of the natural and variable motion of thoughts – as an invaluable tool for carrying out the work of natural philosophy.'[72] Hock takes this further to suggest that Cavendish moves beyond Lucretius, as her 'poetry is not a superficial inducement to natural philosophical insight but rather fundamental to its practice.'[73] This is an important observation. The imaginative exercise of the

---

Sarah E. Moreman, 'Learning their Language: Cavendish's Construction of an Empowering Vitalistic Atomism', *Explorations in Renaissance Culture*, 23 (1997), 129–44.

[71] For further information on the five-part structure of *Poems and Fancies*, see Blake, 'The Interlinked Structure of *Poems and Fancies*' in 'Reading Poems (and Fancies)', <http://library2.utm.utoronto.ca/poemsandfancies/introduction-to-cavendishs-poems-and-fancies/#intro-2-sec-1> [accessed 26 Feb. 2021].

[72] Hock, 'Fanciful Poetics', 772.

[73] Ibid., 801.

atom poems was instrumental to establishing a methodology that Cavendish would maintain for the rest of her scientific and literary career. These poems come in different shapes and sizes, engaging in metatextual play with their subject matter. Like atoms, the poems create anew and build new worlds; they stand alone as individual entities, but also combine to empower Cavendish's philosophies of 'fancy' and of thought. Constantly shifting in shape as they move between transient forms, her sentient atoms provide a voice both for the emerging author and for the material world of which they are a part, thus enabling Cavendish's theory of individual perception belonging to all material things.

This emerges gradually over the course of the first section, as the atom poems gather to create, in the words of Hero Chalmers, 'the impression of a panoply of images passing rapidly before the reader's eye'.[74] Cavendish establishes the importance of fancy by introducing the superiority of its atomic components. 'Of Loose Atoms' suggests:

> In every brain there do loose atoms lie,
> Those which are sharp, from them do fancies fly.
> Long airy atoms nimble are, and free,
> But atoms round and square are dull and sleepy.

The reader must engage with the act of fancy to explore the natural philosophical content of the poem. 'Of Loose Atoms' is dedicated to the subject of things invisible, which must be imagined to be conceived; additionally, the poem does not join the dots between its imaginative observations, thereby leaving the reader to form her own connections and conclusions. Atoms in the brain are depicted as 'loose' and yet they 'lie', which contrastingly suggests stillness; the next line picks up the pace by claiming that some are sharp, and 'from them do fancies fly'. There appears to be a distinction between the sharp atoms and the fancies that *from* them do fly, but this is left unexplained and is complicated further by the 'Long airy

---

[74] Hero Chalmers, '"Flattering Division": Margaret Cavendish's Poetics of Variety', in *Authorial Conquests: Essays on Genre in the Writings of Margaret Cavendish*, ed. Line Cottegnies and Nancy Weitz (London, 2003), pp. 123–44 (pp. 127–28).

atoms' of the following line. These 'nimble' atoms may be the same as those that produce fancies, or they may be the substance of the fancies themselves, but Cavendish resists confirming one way or the other. The allusion to these atoms as 'free' was a textual amendment in the 1664 and 1668 editions to what was printed originally in 1653, that '[t]hose that are long, and Aiery, nimble be'.[75] Cavendish's alteration to the line heightens the dexterity of fancy, which flies from subject to subject and can entertain different perspectives and possibilities of reality, a key motivation of her poetry and her natural philosophy.

She continues her focus on the vitality of sharp atoms – the elements of fancy – in 'What Atoms Make Life', where she writes, commenting on the different shapes of particles:

> Those which are pointed, straight, quick motion give,
> But those that bow and bend, more dull do live.
> For life lives dull or merrily
> According as sharp atoms be.

Cavendish's 'pointed, straight' atoms are like the sharp, 'loose atoms' of the brain. She accentuates the desirability of 'quick motion' by making the leap, between the sixth and seventh lines, from discussing the lives of the atoms themselves to expanding her focus to 'life' in general. This is one of many moments in the poems where Cavendish asserts that living things depend in turn on the lives of atoms for their wellbeing. Here, she repeats the adjective 'dull' to associate the sluggish heaviness of bent, immobile atoms with slowness of understanding and listlessness. On the contrary, life lives 'merrily' when atoms are sharp. Appropriately for a collection called *Poems and Fancies*, Cavendish's choice of adverb gestures at a life of pleasure and exuberance but also suggests wit and cleverness (a significance that corresponds with the additional meaning of 'sharp' as quick-witted.) In 'Of Sharp Atoms', Cavendish confirms that

---

[75] See Blake, 'Of Loose Atoms' in *Poems and Fancies*, <http://library2.utm.utoronto.ca/poemsandfancies/2017/06/09/of-loose-atoms> [accessed 26 Feb. 2021].

> Motion the sharpest atoms doth mount high,
> And like to arrows swift, doth make them fly.
> And being sharp and swift, they pierce so deep,
> As they pass through all atoms as they meet.

The motion described above could just as easily apply to the faculty of fancy as it is exercised throughout the atom poems. Sharp atoms, the elements of the imagination, fly 'high' and pierce 'deep'; they have access to higher and more obscure areas of reality. Cavendish forms a simile with 'arrows swift' that stresses the penetrability and speed of sharp atoms, but also the potential danger of their ability to 'pierce' and to 'pass through all atoms', scattering the components of forms in their wake. It may have been to separate imaginative acts from the tensions of this threat that 'Of Loose Atoms' explains fancy flies *from* sharp particles. As the preface to *Natures Pictures* confirms, Cavendish's fancy is constructive, not destructive: it is intimate with the inner workings of nature through its double role as a creator and as 'a naturall Creation'.

The opening poems of *Poems and Fancies* are both the founding atoms of Cavendish's literary career and the spaces in which she first demonstrates the natural philosophical productivity of fancy. It is no coincidence that the beginning of the collection sets the imagination free to explore the nature of atoms, a subject that both augments the suitability of fancy as scientific method and, in explaining where it comes from, shows its infinite appetite for variety and uncovering possible truths. By the end of Part I, Cavendish has explored the multifarious particles of life and fancy to the extent that, in the closing poems, her creative investigations can 'pierce so deep' as to imagine the possibility of hidden microworlds. 'Of Many Worlds in this World' marvels:

> Although they are not subject to our sense,
> A world may be no bigger than twopence.
> Nature is curious, and such works may shape
> Which our dull senses easily escape.

The act of fancy developed in the atom poems creates opportunities for deeper exploration, even while maintaining a scepticism about what can and cannot be known and remaining in the mood of possibility ('A world *may*'; 'such works *may* shape'). Possibilities open enquiries while premature conclusions, which rest on the restricted abilities of the 'dull senses', shut them down. Cavendish's use of 'dull' again denotes slowness of understanding (she repeats the adjective from her previous poem in the collection, 'It is Hard to Believe that there Are Other Worlds in this World', which similarly admits 'many things our senses dull may scape, / For they're too gross to know each form and shape'). 'Dull' also declares sticking to the senses uninteresting and no match for 'curious' Nature, who is subtle and skilful in her workings. It is fancy that builds and nurtures an authorial intimacy with the natural world that makes further investigation possible – and it is the opening exploration of the atom poems that inspires Cavendish to imagine the infinite variety of nature.

Critical readings have tended to associate Cavendish's atom poems with an early stage of her thinking that she declined to develop in her later, mature philosophy.[76] 'Cavendish's early atomism' is commonly contrasted with her subsequent disapproval of experimental science and desire for a caring, maternal nature.[77] This view, however, falls short of recognising the key factors that

---

[76] Critics have remained undecided on the influence Cavendish's early atomism came to bear on her philosophical career. Karen Detlefsen, for example, argues that Cavendish cannot persist with atomic theory because she is against the atomistic behaviour of individuals; her mature philosophy rests on a plenum because it 'conceives of the non-human world in terms of the human, social world, which is explanatorily primary for her' (p. 202). Detlefsen argues against Stephen Clucas and Jay Stevenson, who both claim Cavendish never entirely leaves atomism behind. Clucas's argument – that what Cavendish objected to was 'mechanical atomism', leaving her the space to develop a 'vitalist atomism', if there can be such a thing – is especially convincing. Not enough has been said on how atomism shaped – and endured in – Cavendish's imaginative fiction, and how this close partnership between atoms and 'Fancy' in turn affected Cavendish's natural philosophy. Clucas, 'The Atomism of the Cavendish Circle'; Stevenson, 'The Mechanist–Vitalist Soul of Margaret Cavendish', *Studies in English Literature 1500–1900*, 36 (1996), 527–43.

[77] For example, in Lisa T. Sarasohn, *The Natural Philosophy of Margaret Cavendish*; Anna Battigelli, *Margaret Cavendish and the Exiles of the Mind* (Lexington, 1998), p. 49, and Deborah Boyle, 'Margaret Cavendish's

attracted Cavendish to atomism in the first place. She designs her atom poems to nurture her thoughts and her text in ways that have been much overlooked. It is in her epistle to her brother-in-law Charles Cavendish, atomist and former member of the Northumberland Circle, that she offers the following defence:

> spinning with the fingers is more proper to our sex than studying or writing poetry, which is the spinning with the brain. But I, having no skill in the art of the first (and if I had, I had no hopes of gaining so much as to make me a garment to keep me from the cold), I made my delight in the latter, since all brains work naturally, and incessantly, in some kind or other, which made me endeavor to spin a garment of memory to lap up my name, that it might grow to after ages.

Cavendish uses her professed ignorance in spinning to argue for the exercise of textual composition in its place.[78] In so doing, she uses the commonplace image of female industriousness to defend her reasons for writing poetry, wittily twisting the image of women's sewing practices to conclude that her needlework is in her authorship. If she cannot make 'a garment' out of the more conventional occupation, she can through her verse. She endeavours 'to spin a garment of memory to lap up my name, that it might grow to after ages': 'lap' describes both the permeability of the textual garment that absorbs the writer's identity and the act of enwrapping her name to give it vitalistic form.[79] Cavendish stores her ideas in text to prepare for an afterlife in the minds of a readership with superior judgements, and the form with which she chooses to begin her literary career is the atom poem. If writing poetry is 'spinning with the brain', its materials are the nimble, sharp atoms of fancy.[80]

Nonfeminist Natural Philosophy', *Configurations*, 12.2 (2004), 195–227 (197–98).

[78] She refers at several points throughout the *Sociable Letters* to those who have criticised her negligence of household duties; the letter 'On Her Housewifery' confronts accusations of 'letting her Servants be Idle without Employment' with the defence that 'none can want Employment, as long as there are Books to be Read' (p. 313).

[79] I am grateful to James Jiang for pointing out the significance of 'lap' as an act of enfolding, binding or clothing (see *OED* for 'lap, v.2'.)

[80] Cavendish confirms moreover in 'It is Hard to Believe that there Are Other Worlds in this World', towards the end of Part I, that 'fancy cannot be

A complex relationship emerges between Cavendish's endeavour to 'spin' an enduring garment out of her texts and her commitment to embracing multiple theories as the nearest depiction of truth. Even with her love of variety and belief in the productivity of ambiguity, there is the need for verbal expression to deliver its possibilities. A consequence of the open flexibility of her 'fancy' is an anxiety over the accuracy of form and communication in preserving plural observations. This is mirrored by the motions of particles in her atom poems, which, in spite of their innate curiosity and endless opportunities for contributing to creation, ultimately are only productive if they come together to take form. Whether in the context of the natural world or in Cavendish's collection of poems, atoms need to take 'their right places' to provide subjects for the imagination.[81] Likewise, the communication of thoughts and fancies requires that they be set in some form that exhibits their content. In another of the introductory letters to *Poems and Fancies*, the 'Epistle to Mistress Toppe', Cavendish defends her literary composition again by appeal to the conventional occupation – or lack of, in this case – of her gender:

> For the truth is, our sex hath so much waste time, having but little employments, which makes our thoughts run wildly about, having nothing to fix them upon, which wild thoughts do not only produce unprofitable, but indiscreet actions, winding up the thread of our lives in snarls on unsound bottoms. And since all times must be spent either ill, or well, or indifferent, I thought this was the most harmless pastime.

---

without some brains'.

[81] I quote from 'All Things Last or Dissolve According to the Composure of Atoms', which explains the dissolution of bodies formed from 'loose atoms': 'For motion's power tosseth them about, / Keeps them from their right places: so life goes out.' The importance of atoms taking their correct places is a running theme through the poems, usually manifesting at points of anxiety about threats to life. For example: in 'What Atoms Make Change', forms dissolve when atoms 'did not their right places take'; in 'What Atoms Make the Wind Colic', illness occurs when atoms, 'being forced, not in the right places lie'; in 'Atoms and Motion Fall out', 'Motion lets them not their places take'.

Her argument strikes a similar chord to the claim proposed in 'To Natural Philosophers', not entirely convincingly, that 'fiction' is for 'pastime'. 'To Natural Philosophers' makes it clear that her intended audience is those who fall into this social trap of idleness: 'For I had nothing to do when I wrote it, and I suppose those have nothing or little else to do that read it.' In her correspondence with Mistress Toppe, Cavendish describes the thoughts experienced by women without employment running 'wildly about, having nothing to fix them upon', as though they were her 'loose atoms' in motion. Her wording brings female cognitive activity into close contact with atomic movement. As atoms seek to rest and create new forms, Cavendish's thoughts – and her writings in turn – similarly desire to focus profitably on their subjects. It is indubitable that she later gave preference to vitalism as her chosen philosophy of nature, but Cavendish's combined exploration of poetical and scientific interests in *Poems and Fancies* came from an empathetic association between female authorial self and atomised matter: hers was a *vitalist* atomism, which needed to create forms to pave the way for her future natural philosophical developments.[82]

Cavendish's poetic voice is consistently familiar with her atoms, their motions and their experiences. Her atoms, albeit inspired by Lucretius and what she had heard at second hand of Pierre Gassendi and Thomas Harriot, are no senseless, reductionist components of mechanised matter.[83] As Lisa Sarasohn has pointed out, the minute particles that both configure and populate her worlds are unmistakably, and rather controversially, alive.[84] Even before she progresses to embrace the hypothesis that

---

[82] See also Stevenson, 'Mechanist–Vitalist Soul'. John Rogers focuses on Cavendish's female-centric vitalism in *The Matter of Revolution*, pp. 177–211.

[83] B. J. Sokol has argued that Cavendish became aware of the atomist and mathematician Thomas Harriot through Charles Cavendish, who transcribed many of his mathematical manuscripts on the nature of infinity ('De Infinitis'). For more information, see B. J. Sokol, 'Margaret Cavendish's *Poems and Fancies* and Thomas Harriot's Treatise on Infinity', in *A Princely Brave Woman: Essays on Margaret Cavendish*, ed. Stephen Clucas (Farnham, 2003), pp. 156–70 (especially pp. 159–60).

[84] Sarasohn, p. 43.

a 'world' may take form in an 'earring', and to merge 'atoms in the brain' with fairies, Cavendish invests her elements with feeling and agency:

> 'Tis several figured atoms that make change,
> When several bodies meet as they do range.
> For if they sympathize and do agree,
> They join together, as one body be.
> But if they meet, like to a rabble rout,
> Without all order running in and out,
> Then disproportionable things they make,
> Because they did not their right places take.[85]

The above poem, 'What Atoms Make Change', plays with a number of commonplaces from Lucretius and early modern 'allegories of the atom' – atoms dance and make figures; atoms fall out of place and into chaos.[86] The associations may to some extent be conventional, but Cavendish's take on them most certainly is not. Her particles congregate in a dance that is not primarily the dance of atoms in disorder, but a sequence of movements which, if followed exactly, result in a unified, coherent 'body'. There are 'right' and wrong places to assume, and these seem not merely to be left to chance but down to the opinions and sympathies of the individual atoms themselves. Thus Cavendish plays, suggestively, with 'figure': 'figured atoms' follow set, pre-choreographed movements – this gestures back to the possibility that atoms might be 'predestined' to work a particular 'Fate', but as in her earlier poem, predestination does not entail certainty or command. What 'may' be in 'A World Made by Atoms' comes to pass 'if' chance encounters inspire particular responses and lead to one of multiple possible outcomes. Cavendish recognises a set of possibilities in the movements of atomic matter, all of which rely upon the choices of anthropomorphised atoms and are encapsulated in the ambiguity of the resulting

---

[85] See 'Of Fairies in the Brain', Clasp IV–V, *Poems and Fancies*. Cavendish's 'A World in an Earring' is in Part I, directly after 'Of Many Worlds in this World'.

[86] Reid Barbour refers to 'allegories of the atom' in *English Epicures and Stoics: Ancient Legacies in Early Stuart Culture* (Amherst, 1998), p. 35.

figure: as dance, emblem, poetic form – especially the closure of rhyming couplet – or written character. The success of her poem and indeed the entire collection, not to mention the world she annotates, relies upon whether her figures do or do not 'their right places take'.

The importance Cavendish bestows the placement of atoms is a running theme throughout her collection. Her poem 'All Things Last or Dissolve According to the Composure of Atoms' is highly suggestive of the close affinity between her corpuscular theory and the actualisation of her literary ambition:

> Atoms which loosely join do not remain
> So long as those which closeness do maintain.
> Those make all things i'th'world to ebb and flow,
> According as the moving atoms go.
> Others in bodies, they do join so close,
> As in long time, they never stir nor loose.
> And some will join so close and knit so fast,
> As if unstirred they would forever last.
> […]
> Those bodies where loose atoms most move in,
> Are soft and porous, and many times thin;
> Those porous bodies never do live long.
> Why so? Loose atoms never can be strong.
> For motion's power tosseth them about,
> Keeps them from their right places: so life goes out.

For Cavendish, who desires to author a work that will 'grow to after ages', there is but a fine line between the textual and the atomic modes of composition. Her text, figured as a garment, could be said to follow its atomic principles in longing to maintain a 'closeness' that will 'knit so fast' so that its contents 'would forever last'. She writes in an earlier poem, 'The Joining of Several Figured Atoms Make Other Figures':

> When several figured atoms well agreeing
> Do join, they give another figure being.
> For as those figures join in several ways,
> So they the fabric of each creature raise.

The growing series of atom poems, if 'well agreeing', connect to knit together a 'fabric' that is both a 'garment of memory' and an immortal vehicle for the author. Cavendish's atoms, and consequentially her poems, have the power to give or to remove her authorial life and repute.

Cavendish's atom poems double as the foundational atoms of her natural philosophical and literary career. As forms, they contract – not unlike the 'indivisible' atom – her two most significant and defining interests as a writer, natural philosophy and poetry, into a single space. It was of vital importance to Cavendish what places these poems took. This sense of right and wrong extended into her editorial activity: once she had sent a book off to the printers, Cavendish rarely left it alone, frequently suggesting restructures and rewrites that took works into two, and sometimes even three, editions. Her poetical and imaginative works were subject to greater revision than her natural philosophical essays. As with her vitalist attitude towards philosophical thought, Cavendish was flexible in her verbal expression and worked with her text as something live and changeable. She was apt to change the method of representation if, as the Dutch experimentalist Constantijn Huyjens once commented in their correspondence, she might 'make choice of a better' on another day.[87] James Fitzmaurice has made the claim for the high level of manuscript authorial corrections in *Sociable*

---

[87] Cavendish's emphasis on allowing changeability of philosophical thought made a distinct impact upon the Dutch natural philosopher Constantijn Huygens, who corresponded with Cavendish over a number of years on the subject of his experiments with glass. On discovering a conflict between her scientific opinion and his laboratory work, Huygens asked the Duchess to suggest ways in which he might move forward in his enquiries, flattering her 'most ingenious perspicacitie':

> I remember yr Excellency would declare unto me that beeing to day of an opinion in matter of Philosophie, she would not to be bound to it that to morrow she might not make choice of a better.

Quoted from Nadine Akkerman and Marguérite Corporaal, 'Mad Science Beyond Flattery: The Correspondence of Margaret Cavendish and Constantijn Huygens', in *Ashgate Critical Essays on Women Writers in England 1550–1700: Vol. 7, Margaret Cavendish*, ed. Sara H. Mendelsohn (Farnham, 2009), pp. 263–304 (p. 276).

Letters (1664), *The Life of the Thrice Noble, High and Puissant Prince William Cavendishe* (1667) and *Playes: Never before Printed* (1668).[88] Liza Blake has shown moreover that Cavendish was engaged with editing concerns long before the Restoration.[89] *Poems and Fancies* underwent three editions over the course of her lifetime, during which the author reorganised its contents, altered words in lines of poetry and provided first more and then less expository information in the margins. What appeared first as *Poems, and fancies* in 1653 was renamed *Poems, and phancies* in 1664. The change in title is likely to have been the decision of the printer, rather than the author. Regardless of its origin, the decision to alternate between 'fancy' and 'phancy' may have been grounded on supplying a more attractive and enticing title. Both words are identical in sound and etymology and were nearly interchangeable within early modern discourse and literature, but there was subtle variation in their respective significance. The term 'fancy' was undergoing a transition during this period towards its predominant modern sense of 'caprice, whim, fanciful invention', while 'phancy' held the more credible connotations of synonymy with 'imagination' and 'visionary notion', as derived from scholastic psychology.[90] By 1668, however, the third edition of the work was published as *Poems, or, Several fancies in verse, with the Animal Parliament in Prose*. The decision to change back may have been for the simple reason that 'Fancies' sold better and looked less outdated, or it may again have been the printer's preference – Cavendish changed publishers between each of the three editions.[91]

---

[88] Many of the works corrected by hand were given by the author as presentation copies to libraries in Oxford and Cambridge. See James Fitzmaurice, 'Margaret Cavendish on Her Own Writing: Evidence from Revision and Handmade Correction', *Papers of the Bibliographical Society of America*, 85 (1991), 297–397 (299–300).

[89] Liza Blake, 'Pounced Corrections in Oxford Copies of Cavendish's Philosophical and Physical Opinions; or, Margaret Cavendish's Glitter Pen', *New College Notes*, 10 (2018), no. 6 <https://www.new.ox.ac.uk/node/1804> [accessed 26 Feb. 2021]. See also Blake, 'Changes Between First and Second Editions', in 'Textual and Editorial Introduction', *Poems and Fancies*.

[90] See the definitions of 'Fancy, n. and adj.', *OED*.

[91] *Poems, and fancies* (1653) was published by J. Martin and J. Allestrye; *Poems, and phancies* (1664) by William Wilson, and *Poems, or, several Fancies in*

Poems and Phancies is advertised on its frontispiece as a work that has been 'much Altered and Corrected'. The greatest change Cavendish imposed upon her material was a reorganisation of its contents, most notably by reordering the series of the 'atom' poems at the opening of the volume. When *Poems and Fancies* was printed in 1653, Cavendish oversaw her publication from the distanced position of exile in Antwerp and was unable to intervene at the latter stages of the process.[92] A British Library copy of the original edition contains a plea, handwritten by the author on the final page of prefatory material:

> reader let me intreat you to consider only the fancyes in this my booke of poems and not the languede numbers nor rimes nor fals printing for if you doe you will be my condeming iudg which will grive my mind.[93]

In the first edition the presentation of the opening natural philosophical poems is as atomistic as the subject matter, but the restructured contents of the 1664 edition reveal that Cavendish had intended to group her poems more cohesively. The care she took reveals, in spite of critical views to the contrary, that she had continued to rethink and develop her thoughts on atomism into the Restoration. Even if the editorial interventions were only carrying out what was originally intended in the early 1650s, the changes were considered important enough to be put into practice more than ten years later. As Emma Rees has pointed out, the philosophical investigations of Cavendish's poems correspond with the major themes of Lucretius's *De Rerum Natura*: the nature of being and the vacuum, atoms, motion, infinity, mind and perception, vision, the weather and disease feature prominently in both works. While Rees acknowledges that Lucretius provided 'form and subject matter' fundamental to her first published work', the structure of *De Rerum Natura* may also have influenced how

---

Verse: with the Animal Parliament, in Prose (1668) by A. Maxwell.

[92] See Fitzmaurice, p. 299. Katie Whitaker discusses this in detail in *Mad Madge*, where she argues a publication date in the early months of 1653 was likely, before Cavendish travelled to Antwerp to join her husband. Whitaker, p. 155. See also Liza Blake, 'Textual and Editorial Introduction'.

[93] The copy is British Library, C 39. h. 27(1).

Cavendish reordered her poems for the 1664 edition.[94] *Poems and Phancies* opens with verses on atoms in each of the four elements, air, earth, fire and water, before proceeding organically to animal and vegetative matter, life and death, sickness, motion, the sea, the sun and meteorology. Lucretius also begins with a description of the atom and of the nature of matter, followed by studies of motion and mind, death, perception, and meteorological phenomena. It is highly likely that Cavendish reorganised her atom verses with the structure of the influential Epicurean poem in mind.

Far from fleeting and insubstantial, Cavendish's atom poems take their places in *Poems and Fancies* to secure their author's legacy. They provide an initial base for the generation and gathering of 'Fancies', but – as the evidence of Cavendish's sustained authorial intervention suggests – they continued to form and re-form over succeeding years. In the prefatory material of her first publication, Cavendish linked her career-defining atom poems with the free movement of her thoughts as female writer. She would either be 'a world, or nothing'. By writing on atoms, the closest entities to 'nothing' in creation, Cavendish could wryly pass off her poems as inconsequential, harmless frivolities. While doing so, however, she could build literary structures to create a garment that would 'lap' up her name, 'that it might grow to after ages'.

The atom poems began the process of creating a new world for Cavendish and her female readers. They offered escapism, but there was nothing insubstantial to their fictional reality. Cavendish remained committed throughout her writings to the importance of imaginative world-making as a woman's right. The most well-known assertion of this comes from *The Blazing World*, in which a sensational kidnap story morphs into an account of how an anonymous 'Lady' becomes Empress, and natural-philosopher-in-chief, of a new world. In the second half of the fantastical tale, the new Empress seeks to employ a scribe and ends up appointing the services of the Duchess of Newcastle herself (somewhat ironically

---

[94] Emma L. Rees, *Margaret Cavendish: Gender, Genre, Exile* (Manchester, 2003), p. 70. More recently, Jessie Hock has argued that *Poems and Fancies* 'versifies the atomist underpinnings of the cosmos in imitation of Lucretius's great first-century BCE atomist poem'. Hock, 767.

given the original's appalling handwriting).[95] The consequent conversations and adventures between the two women personify the free and unbounded quality of 'Fancy' in the mind – the characters are liberated to explore and create multiple worlds. In a discussion with spirits in the sky of the Blazing World, the Empress is advised to create an original world within her mind rather than seek to conquer heavier pre-existing matter. The spirits explain that

> every human Creature can create an Immaterial World fully inhabited by Immaterial Creatures, and populous of Immaterial subjects, such as we are, and all this within the compass of the head or scull; nay, not onely so, but he may create a World of what fashion and Government he will, and give the Creatures thereof such motions, figures, forms, colours, perceptions, &c. as he pleases, and make Whirl-pools, Lights, Pressures, and Reactions, &c. as he thinks best; nay, he may make a World full of Veins, Muscles, and Nerves, and all these to move by one jolt or stroke: also he may alter that World as often as he pleases, or change it from a Natural World, to an Artificial; he may make a World of Ideas, a World of Atoms, a World of Lights, or whatsoever his Fancy leads him to. And since it is in your power to create such a World, What need you to venture life, reputation and tranquility, to conquer a gross material World?[96]

Though the spirits claim original worlds are 'Immaterial', they are only 'Immaterial' in the sense that they are separated from sensitive and inanimate matter. Immateriality is incompatible with Cavendish's philosophy but referenced here because the spirits actively satirise Henry More.[97] Despite this satirical purpose, there are

---

[95] In her letters dating from her courtship with William, Cavendish frequently apologised for her terrible handwriting: one of her letters signs off with the somewhat endearing excuse, 'if you cannot reed this leter blam me not for it was so early I was half a slep'; another begs the recipient to 'lay the falt of my wrighting to my pen'. British Library, Additional MS 70499, fol. 267 and fol. 259b.

[96] *The Blazing World*, 'The Description of a New World', p. 97.

[97] As Sarah Hutton has argued, 'the purpose of the allusions to More in *Blazing World* is to refute his philosophy. It is with consummate irony that Cavendish has her Empress consult spirits, whose very existence as incorporeal beings is impossible according to her own philosophy.' Hutton, 'Margaret Cavendish and Henry More', in *A Princely Brave Woman*, pp. 185–98 (p. 194).

multiple ways of reading this passage – as is fancy's wont – and the spirits' advice does deliver an important message about the creative power of the imagination. As the asyndetic lists of possible qualities that can be changed at a 'jolt or stroke' suggest, worlds in the mind are formed of the more refined, rational degree of matter and are thence infinitely malleable. Whatever the 'fancy' construes – atoms, veins, muscles – is complicit with corporeal reality. This is the other side to a sceptical philosophy where nothing is certain: anything that can be imagined might be. The world in the mind, therefore, is no utopia – or 'no-place' – but a superior place. Cavendish asserts the superiority of subjective experience, but also emphasises that the recreation of re-creation enables fuller philosophical understanding of the natural world. It is no coincidence that the single question the Duchess of Newcastle is asked by friends on return from the Blazing World is 'what Pastimes and *Recreations* her Majesty did most delight in?' (italics mine).[98]

'A WORLD OF ATOMS'

In a philosophy where any given thought can reshape the universe, creative fiction acquires the power to re-write the world. Cavendish was aware that 'A World of Atoms' constructed in the mind was no false fiction, but a generative, creative 'fancy' bearing real potential for change. Atomically, and according to Cavendish's material philosophy, the 'World' of the mind both was and was not the realm of the imagination; there is little to distinguish between the imagined atoms of the fancy and the material atoms composing brain and thought. Both of these atomic creations (imagined and of the material self) are, to adopt a phrase recently used by Claire Preston, 'true fictions' – not merely in the role of 'similitudes that illuminate discovery', but in the sense that they are, and yet are not, there.[99] They are promises and prophecies; glimpses into future possibilities and traces from historical movements. What is imagined must be conceived, in some form: the distinction between

---

[98] Ibid., 'The second part of the description of the New Blazing World', p. 32.
[99] Claire Preston, *The Poetics of Scientific Investigation in Seventeenth-Century England* (Oxford, 2015), p. 5.

physical and figurative atoms, between particles in the brain and particles in the imagination, breaks down under the pressure of the atomic hypothesis underlying all created beings and acts.

Hence, the quotations from Cavendish and Pulter's work that opened this chapter: '…by other atoms thrust and hurled / We give a *being* to another world'; 'Small atoms of *themselves* a world may make…' (italics mine). The poetical atoms of both writers are associated with self-identification and motivation. For Pulter, atoms have 'being'; for Cavendish, they have 'selves' and bequeath power to the self (in this case, the female author) writing at the margins of society. Atoms are world-changing for Cavendish and Pulter. Their authorial priorities admittedly run in different directions – Cavendish deals in the positive construction and amendment of ideas, while Pulter locates her discoveries in lack and cosmic disorder. In her preface 'To the Reader', Cavendish imagines the book of *Poems and Fancies* as her child, an embodied textual promise of her authorial immortality and maturation of intellect and wit.[100] Pulter also draws on the figure of the infant, but associates 'embryon' being with the *necessary* imperfection of all material life. The related 'promise' of immortality in her case, a promise of rebirth at the atomic corruption of the present world, is achieved through trust, humility and acceptance of mortal limitations. But Pulter is not without ambition. Atoms, as Cavendish reminds us, can either make 'nothing', or a 'world'. While Cavendish acts positively as creator in bringing atoms (starting with 'atom poems') together to create new worlds on the page and in the mind, Pulter sticks with what she knows, the pre-created matter of her immediate surroundings, and anticipates atomic dissolution – world-breaking rather than world-making – to find release in 'another world'.

---

[100] Cavendish writes: 'True, you may tax my indiscretion, being so fond of my book as to make it as if it were my child, and striving to show her to the world in hopes some may like her … wherefore pity her youth and tender growth, and rather tax the parent's indiscretion than the child's innocency.' 'To the Reader', <http://library2.utm.utoronto.ca/poemsandfancies/2019/05/04/to-the-reader-2/> [accessed 26 Feb. 2021].

The atoms, poems and worlds of this chapter begin to show the appeal a poetics of the atom could have to those who might, for various reasons, be considered outsiders within seventeenth-century intellectual society. In writing their 'atom poems', Cavendish and Pulter strove to make new and be made new. The accommodative potential of the 'indivisible' atom expanded the opportunities for exercising wit, imagination and philosophical contemplation, and pushed the limits of authorial ambition. Through intimate acquaintance with the atom, the smallest yet most significant of created material things, Cavendish and Pulter 'enfranchised' their verse and gave being to alternative worlds.

Chapter 4

THE ATOM IN GENESIS: LUCY HUTCHINSON'S *ORDER AND DISORDER*

Within early modern intellectual culture, the figure of the first man and the smallest, indivisible particle of the natural world had an important factor in common: they were both first principles. Both Adam and the atom were, in their separate ways, founders of all generation and corruption that succeeded them and, for many seventeenth-century theologians and philosophers, desirable end-objects of knowledge. John Milton was certainly aware of the status held by the figure of Adam in contemporary philosophy. His depiction of the prelapsarian man responds to widespread views of Adam as first philosopher: the first principle of mankind, capable in turn of perceiving the essence of other first principles, with an innate ability to understand the true nature of things. There are two explicit references to 'atoms' in *Paradise Lost*. The first, and most well-known, occurs at the point of Satan's flight through Chaos; the second arises during a conversation between the first man and the archangel Raphael.[1] Milton's Adam combines blessed innocence with the role of the experimental philosopher, a winning

---

[1] The oft-quoted passage from Book 2 describes Satan's view of Chaos, with the following observation: 'For Hot, Cold, Moist, and Dry, four champions fierce / Strive here for mastery, and to battle bring / Their embryon atoms' (2. 898–900). John Milton, *Paradise Lost*, ed. Alistair Fowler (Harlow, 2007). All further quotations from *Paradise Lost* will be taken from this edition and cited in the text by book and line number.

combination, commonly aspired to by seventeenth-century philosophers who longed to restore the same direct acquaintance with the origins of things.[2] Following Raphael's account of the creation, Adam – with awe and gratitude but also with doubt, possibly sparked by the stimulus of recent conversation – replies that the angel:

> largely hast allayed
> The thirst I had of knowledge, and vouchsafed
> This friendly condescension to relate
> Things else by me unsearchable, now heard
> With wonder, but delight, and, as is due,
> With glory attribúted to the high
> Creator; something yet of doubt remains,
> Which only thy solution can resolve.
> When I behold this goodly frame, this world
> Of heaven and earth consisting, and compute
> Their magnitudes, this earth a spot, a grain,
> An atom, with the firmament compared
> And all her numbered stars, that seem to roll
> Spaces incomprehensible (for such
> Their distance argues and their swift return
> Diurnal) merely to officiate light
> Round this opacous earth, this punctual spot,
> One day and night; in all their vast survey
> Useless besides, reasoning I oft admire,
> How nature wise and frugal could commit
> Such disproportions (8. 7–27)

Adam's equation of the world with an 'atom' relies on a vision of unfathomable clarity and focus, which he acknowledges through his desire for 'unsearchable' and 'incomprehensible' objects of

---

[2] For studies on the early modern intellectual desire to reconnect with prelapsarian nature, see Jim Bennett and Scott Mandelbrote, *The Garden, the Ark, the Tower, the Temple: Biblical Metaphors of Knowledge in Early Modern Europe* (Oxford, 1998); Peter Harrison, *The Fall of Man and the Foundations of Science* (Cambridge, 2007); Joanna Picciotto, *Labors of Innocence in Early Modern England* (Cambridge, MA, 2010); Robert N. Watson, *Back to Nature: The Green and the Real in the Late Renaissance* (Philadelphia, 2006).

knowledge. The first man, geographically isolated within the bounds of the garden, has looked up at the skies in order to look down upon and perceive the scale of the earth. His computations expose vertiginous 'disproportions' within the natural universe: 'this earth a spot, a grain, / An atom, with the firmament compared'. Notably, however, Adam's terms for describing his world progress not only in descending order of size or significance, but in ascending order of creative potential. His initial 'spot' could be as meagre as the 'dim spot, / Which men call earth', looked down on (both physically and judgementally) by the Spirit at the opening of Milton's *Comus*, but the following descriptors mark changes in Adam's perspective.[3] More productively, the 'grain', though minute, is an origin of life and the 'atom', smaller still, has the potential to contribute to the creation of all things (for, as Satan recognises in Book 2, atoms are 'embryon') (2. 900). If Adam's understanding of atomism is at all comparable to Satan's, or indeed Milton's, his theories are coloured by Lucretius. When he lifts his head to the dizzying heights above, the solid world beneath his feet becomes as precarious as a single particle in flux. This 'punctual spot' – punctual in the sense of 'point-like', but also regular in its motions – is no longer as fixed as it seems; in comparison to the many 'numbered stars', the world moves and undergoes as many transitions as a 'grain' or an 'atom'.[4]

It is left ambiguous as to how Adam first encountered atomism, and as to whether his above-described 'reasoning' took place habitually in the days following his creation or in inspired response to Raphael's sharing of information. His claim to have 'oft' pursued these thoughts makes the former seem likely. The nature of Adam's atom is left similarly unclear. The seventeenth-century 'atom'

[3] John Milton, *A Masque Presented at Ludlow Castle, 1634 [Comus]*, in *Milton: The Complete Shorter Poems*, ed. John Carey, rev. 2nd edn (Harlow, 2007), pp. 173–233 (p. 180).
[4] Puck Fletcher has discussed the significance of shifting perspectives in *Paradise Lost* through an understanding of relative space, arguing 'The holding open of multiple cosmic possibilities in Milton's poem cultivates an awareness of the relative experience of space, as explored by means of the (differently) limited perspectives of the characters and narrator.' Puck Fletcher, 'Space and Knowledge in Milton and Newton', 1–23 (forthcoming), 2.

could simply refer to a very small thing, a generalised significance suggested here by the vision of 'spot' and 'grain'. *Paradise Lost* nevertheless counts among its many influences the atomist poetry of *De Rerum Natura*, which Milton was likely to have studied during his senior year at Cambridge.[5] While this influence has been acknowledged in past criticism, and recent years have produced a flurry of studies on Milton's vitalist materialism, there has been little attention paid to the very strong possibility of an atomic matter theory in the century's most famous Christian epic.[6] Indivisible atoms were far from banished from Christian accounts of Genesis. Some years following Milton's biblical poem, in 1679, the first five books of an ambitious poetic narrative of creation (which, at its full length, would span the entirety of Genesis) were published anonymously and later attributed to Allen Apsley.[7] The poem turned out not to be

---

[5] Harris Francis Fletcher, *The Intellectual Development of John Milton: Volume II, The Cambridge University Period, 1625–32* (Urbana, 1961), p. 460. See also Philip Hardie, 'The Presence of Lucretius in *Paradise Lost*', *Milton Quarterly*, 29.1 (1995), 13–24 (13).

[6] Studies of Milton's vitalist materialism include Stephen M. Fallon, *Milton among the Philosophers: Poetry and Materialism in Seventeenth Century England* (Ithaca, 1991); John Rogers, *The Matter of Revolution: Science, Poetry and Politics in the Age of Milton* (Ithaca, 1996); *Milton, Materialism and Embodiment: One First Matter All*, ed. Kevin Joseph Donovan and Thomas Festa (Pittsburgh, 2017). Noël Sugimura reinterprets prevailing views of Milton's materialism in her monograph *'Matter of Glorious Trial': Spiritual and Material Substance in* Paradise Lost (New Haven, 2009). In relation to Milton's atoms, Philip Hardie has reflected on the influence of Lucretius on Book 2 of *Paradise Lost* in 'The Presence of Lucretius in *Paradise Lost*'. David Norbrook has also done more to show the deep presence of Lucretius, including Lucretian atoms, in Milton's famous epic. David Norbrook, 'Milton, Lucy Hutchinson, and the Lucretian Sublime', in *The Art of the Sublime*, ed. Nigel Llewellyn and Christine Riding, Tate Research Publication, January 2013, <https://www.tate.org.uk/art/research-publications/the-sublime/david-norbrook-milton-lucy-hutchinson-and-the-lucretian-sublime-r1138669> [accessed Sept. 2013].

[7] The book is *Order and Disorder; or, the World Made and Undone, being Meditations upon the Creation and the Fall as it is Recorded in the Beginning of Genesis* (London, 1679). The full text of *Order and Disorder*, comprised of twenty books on Genesis, is housed at the Beinecke Library, Yale; Osborn MS fb 100.

the work of Apsley at all, but the composition of his sister – the fiercely Republican Lucy Hutchinson. A staunch Calvinist and devotee of the teachings of Independent minister John Owen, it would seem unlikely that Hutchinson's poetical works should bear any stamp of atoms or Lucretian materialism.[8] As though to support this, in the preface to her creation poem Hutchinson went so far as to condemn the philosophy of *De Rerum Natura* as 'blasphemously against God, and bruitishly below the reason of a man'.[9] There were, however, further reasons for such a direct, emphatic rejection of Lucretius preceding her work. Some twenty years previously Hutchinson had produced what may have been the first full English translation of *De Rerum Natura*, and in spite of her protestations – and the ideological contradictions that ensued – the style of her poetic account of Genesis was strongly influenced by Lucretian poetics and matter theories.[10]

Order and Disorder was almost certainly written in response to *Paradise Lost*, which preceded it in print by twelve years. Numerous parallel passages between the two poems suggest that Hutchinson wrote with the aim of reinterpreting the Miltonic vision of creation. Both make emphatic reference to 'Death's Harbinger[s]' at the point of the Fall; both open their poems by invoking the Holy Spirit; both dwell in detail on the plight of Adam and Eve.[11] There are moreover considerable political and personal connections

---

[8] Hutchinson translated some Latin sections from John Owen's *Theologoumena Pantodapa* (Oxford, 1661). See *On the Principles of the Christian Religion, Addressed to her Daughter; and Of Theology. By Mrs. Lucy Hutchinson*, ed. Julius Hutchinson (London, 1817); also, Crawford Gribben, 'John Owen, Lucy Hutchinson and the Experience of Defeat', *The Seventeenth Century*, 30.2 (2015), 179–90.

[9] 'The Preface', *Order and Disorder*. Quoted from Lucy Hutchinson, *Order and Disorder*, ed. David Norbrook (Oxford, 2001), p. 3. All further quotations from *Order and Disorder* will be taken from this edition and cited in the text by canto and line numbers.

[10] The manuscript is BL Add. MS 19333.

[11] Milton's reference to 'Misery / Death's harbinger' opens Book 9 (9. 12–13); Hutchinson details the corporeal plight of Adam and Eve with 'Deaths harbingers waste in each province make, / While thundering terrors man's whole island shake' (4. 235–36).

between the two writers.[12] Hutchinson's dualist understanding of a distinct separation between body and spirit would appear to be very different from the Miltonic monist vitalism, but she posits that human beings understand what they can of spiritual knowledge and grace through physical activity and communication.[13] Her focus in *Order and Disorder* on discovering divine meaning in matter, therefore, warrants critical attention to parallel the flurry of studies on Milton's accommodating practices and his linguistic and physical

---

[12] It is Norbrook who introduced the possibility of an acquaintance between Hutchinson and Milton that has yet to have been developed further; see '*Order and Disorder*: The Poem and its Contexts', in *Order and Disorder*, ed. Norbrook, pp. xii–lii (p. xiv and pp. xix–xx); see also Norbrook, 'John Milton, Lucy Hutchinson, and the Republican Biblical Epic', ed. Mark R. Kelley, Michael Lieb and John T. Shawcross, *Milton and the Grounds of Contention* (Pittsburgh, 2003), pp. 37–63 (p. 40). Hutchinson and Milton both shared the support of Arthur Annesley, Earl of Anglesey, who was Lord Privy Seal between 1673 and 1682. A moderate and liberal Royalist, he entertained figures as diverse as the Wilmots and the Independent minister John Owen. The owner of the greatest private library in England at the time, he also extended his sympathies to dissenting writers in need of patronage. Milton provided him with a copy of *Areopagitica* in 1644 and Hutchinson, seemingly reluctantly and at her benefactor's request, with a manuscript copy of *De Rerum Natura* in 1675; see the note on 'Arthur Annesley's Copy of Areopagitica', *Milton Quarterly*, 9.4 (1975), 128 and Hutchinson's prefatory letter to Annesley from her *De Rerum Natura*, in *The Translation of Lucretius*, I, 5–15. David Norbrook discusses the manuscript evidence of Hutchinson's creation poem in 'Lucy Hutchinson and *Order and Disorder*: The Manuscript Evidence', *English Manuscript Studies 1100–1700*, 9 (2000), 257–92 (259). For further information, see Annabel Patterson and Martin Dzelzainis, 'Marvell and the Earl of Anglesey: A Chapter in the History of Reading', *The Historical Journal*, 44.3 (2001), 703–26; also, Arthur Annesley's diaries, BL Add. MS 18730 and BL Add. MS 480860.

[13] On Hutchinson's dualism, Joad Raymond, for example, comments that 'Hutchinson's dualist universe is severely hierarchical' and that the 'contrast with Milton is profound', in response to the lines from *Order and Disorder*: 'No; for though man partake intelligence / Yet that, being joined to an inferior sense, / Dulled by corporeal vapours, cannot be / Refined enough for angels' company' (3. 295–98). Joad Raymond, *Milton's Angels: The Early Modern Imagination* (Oxford, 2010), p. 180.

representations of the ineffable.[14] Recently, Julie Crawford has taken this concentration on materiality further to encourage a rethinking of Hutchinson's dualism, arguing that, 'like *Paradise Lost*, Hutchinson's *Order and Disorder* is deeply invested in the propagation and extension of divine spirit throughout the entirety of created matter'.[15] Hutchinson's explorations of material and spiritual experience in the poem envelop complexities and contradictions, with vitalist moments accompanying strong dualist statements – and atoms entering the mix.[16] While it has been acknowledged, criticism on Hutchinson has been curiously reluctant to discuss the undeniable influence of *De Rerum Natura* upon *Order and Disorder*.[17] To claim that her theological poetics were inspired by atomic analogies and the movements of 'indivisibles' within creation is *not* to claim that she was an atomist, or that she in any way aligned her beliefs with Epicureanism. Instead, it is to tease out in her writing an understated deepening of empathy with mortal existence, with the experience of oneself and others as *creatures* raised from matter – the understanding and communication of which are enabled by her atomic perspective of the human condition.

[14] For relevant studies, see John Guillory, 'Ithuriel's Spear: History and the Language of Accommodation', in *Poetic Authority: Spencer, Milton and Literary History* (New York, 1983), pp. 146–71; Annabel Patterson, *Milton's Words* (Oxford, 2009); Victoria Silver, *The Predicament of Milton's Irony* (Princeton, 2001); Martin Kuester, *Milton's Prudent Ambiguities: Words and Signs in his Poetry and Prose* (Lanham, 2009).

[15] Julie Crawford, 'Transubstantial Bodies in *Paradise Lost* and *Order and Disorder*', *Journal for Early Modern Cultural Studies*, 19.4 (2019), 75–94 (76).

[16] Crawford comments, for example, in a reading of Book 3 from *Order and Disorder*: 'while Hutchinson's description of "the noblest creature" having "Earth in his members, Heaven in his mind" (3.28) seems dualistic, the poem imagines the divine distribution of bounties "to all the other members" (3.52) of nature's body in a far more capaciously materialistic sense'. Crawford, 81.

[17] Hugh de Quehen, for example, prefaced his edition of the translation with the claim that none 'of Hutchinson's own poems shows the influence of Lucretius', but this reading overlooks the linguistic and thematic influences of *De Rerum Natura* upon Hutchinson's later writings. See *Lucy Hutchinson's De Rerum Natura*, ed. Hugh de Quehen (London, 1996), p. 7.

Milton's Adam reels at the dizzying heights of creation from the point of this 'goodly frame', a solid base that disintegrates into a single 'atom' following his physical and intellectual ruminations. Hutchinson also draws parallels between the first elements of the material world and the first element of mankind, but to different purpose. In *Order and Disorder*, the ties between Adam and atoms are not so much to do with emphasising the prelapsarian state of simplicity and comprehensive knowledge, as they are in *Paradise Lost*, but on the sudden intimacy between fallen physical being and atomised matter. While Milton's Adam wonders at natural 'disproportions' from the advanced viewings permitted by perfect capacity, Hutchinson's first man suffers physical imbalance at the other extreme, in the wake of his postlapsarian loss of connection with the surrounding world. There is nevertheless another, more positive side to this bond between atomic matter and fallen humanity. Responding to popular contemporary associations between atoms and dust, Hutchinson suggests that mortals can discover soteriological promise in material life: not by prying into the secrets of nature, but by accepting that one's body will sooner or later crumble into atoms – and by trusting that God will reunite the irreducible atoms of the elect.[18]

Parallels between her translation of Lucretius and her later creation poem have been discussed in existing Hutchinson scholarship, albeit briefly. David Norbrook, the leading scholarly authority on Hutchinson's works, explored in a Tate Papers article how *Order and Disorder* 'still reworks lines from her Lucretius' and suggested ideological sympathies in her writing between anticlerical

---

[18] See also Jessie Hock, who has identified a 'Lucretian psychology' in *Order and Disorder*:

> although Lucretian ideas and images generally operate as negative – because mortal and fallen – counterpoints to divine salvation and heavenly pleasures, Lucretian psychology is an essential component of Hutchinson's vision of a Christian universe. While any good Christian seeks to eventually overcome the mortal condition, it is not a stage you can skip.

Jessie Hock, *The Erotics of Materialism: Lucretius and Early Modern Poetics* (Philadelphia, 2020), p. 135.

Republicanism and Lucretian ethics.[19] Most recently, Jessie Hock has encouraged a rethinking of the influence of *De Rerum Natura*'s poetics on Hutchinson's later works, arguing that 'Lucretian desire plays an important part in her elegies and her biblical epic ... as a figure for painful, yet unavoidable, human desire.'[20] Critics have been more cautious when it comes to observing echoes of Lucretian materialism in the post-Restoration poem. Reid Barbour places Hutchinson's translation within the context of 'relations between atomic science and Christian theology at mid century'. He observes, importantly, that there was 'a campaign in the 40s and 50s aimed at reconciling ... atomism to the faith' and argues that 'the study of atoms was seen as a pastoral escape from the religious and political turmoil of the day'.[21] Of Hutchinson's work produced following the Restoration, however, he claims that '[w]hatever uses she may have derived from her work on Lucretius are largely cancelled.'[22] Where previous criticism has drawn attention to the influence of Lucretian materialism on *Order and Disorder*, it has been on the strong parallels between Hutchinson's depiction of a disordered world and the episodes of atomised upheaval outlined in her *De Rerum Natura*. Elizabeth Scott-Baumann has commented on how Hutchinson came 'to see *De rerum natura* as the representation of a world in

[19] In 'Milton, Lucy Hutchinson, and the Lucretian Sublime' (2013), Norbrook shows how the wording of Hutchinson's translation introduces a Lucretius who is an opponent of 'superstition', not religion. On the topic of political sympathy he writes:

> This Puritan interest in Lucretius appears less incongruous in the light of the alliances formed in the 1640s and 1650s between varying groups which, after the collapse of the Church of England, feared the possible establishment of a persecuting Presbyterian Church. The Independent group to which Milton and Hutchinson belonged was ready to ally itself with the absolutist and caustically anti-clerical Thomas Hobbes, and internationally to appeal to traditions of freedom of thought of all kinds.

[20] Hock, *The Erotics of Materialism*, p. 26.
[21] Reid Barbour, 'Between Atoms and the Spirit: Lucy Hutchinson's Translation of Lucretius', *Renaissance Papers* (1994), 1-16 (2-7).
[22] Reid Barbour, 'Lucy Hutchinson and the Atheist Dog' in *Women, Science and Medicine 1500-1700*, ed. L. Hunter and S. Hutton (Thrupp, 1997), pp. 122-37 (p. 123).

decay, as would the world of *Order and Disorder* be without the Christian God at its centre'.[23] It is at such moments that the Earth disintegrates into chaos, as at the threat of 'unlike' elements meeting in unnatural union:

> ... if with their unlike they attempt to mix,
> Their rude congressions everything unfix;
> Darkness again invades the troubled skies,
> Earth trembling under angry heaven lies (*Order and Disorder*, 3. 277–80)

As Norbrook noted in his edition of *Order and Disorder*, Hutchinson uses the word 'congressions' five times in her translation of Lucretius; she also specifically associates it with the collision of atoms in her marginalia.[24]

I have written elsewhere of Hutchinson's Lucretian portrayal of postlapsarian disorder.[25] In their immediate sufferings post-Fall, her Adam and Eve discover the new vulnerability of their flesh through language that merges their physical deterioration with the vocabulary of atomic penetration and dissolution. As Adam and Eve dissolve in sin, their bodies no longer retain independent shape amongst the natural objects of their surroundings. They begin to disintegrate while other substances pierce through their skin, as in the description of their futile clothes devised from foliage:

> But ah! these coverings were too slight and thin
> To ward their shame off, or to keep out sin,
> Or the keen air's quick-piercing shafts, which through
> Both leaves and pores into the bowels flew. (4. 251–54)[26]

---

[23] Elizabeth Scott-Baumann, 'Lucy Hutchinson, Gender and Poetic Form', *The Seventeenth Century*, 30.2 (2015), 265–84 (269).
[24] See also Baumann's observation of this, ibid., 270–71.
[25] Cassandra Gorman, 'Lucy Hutchinson, Lucretius and Soteriological Materialism', *The Seventeenth Century*, 28.3 (2013), 293–309.
[26] A similar image is used by Margaret Cavendish in her poem 'Of Sharpe Atoms':
> Like Arrowes sharpe, Motion doth make them flye.
> And being sharpe and swift, they peirce so deep.

Margaret Cavendish, *Poems, and fancies* (London, 1653), p. 10.

The 'quick-piercing shafts' of the air she describes in *Order and Disorder* recall the motion of Lucretian atoms, moving through the material world and restructuring corporeal forms 'by impulsiue force', 'perplext agitations' and 'secret tumults' (to use the words of Hutchinson's translation).[27] Lucretius regularly applied combative metaphors, like those quoted above, to describe atomic motion. In her corporeal narration of the Fall, Hutchinson draws on the Lucretian recognition of transient physical form. According to *De Rerum Natura* and in Hutchinson's own words, a body disintegrates

> When tis assaulted with a stronger blow
> Then nature can resist, whence there arrives
> A generall tumult in the sence, which driues
> The principles away, destroys their site,
> And doth all vitall motions impede,
> Chacing all matter through the arteries,
> Whose strong concussions, lifes fraile knotts vnties,
> And through the pores eiecting the weake soule,
> Quick dissolution falls vpon the whole. (2. 959–67)

She precedes the abrupt statement 'lifes fraile knotts vnties' with a disruptive caesura; the effect is similar to the disturbance of poetical order during the Fall in her biblical poem, where the manipulation of 'thundering terrors' increases the syllable count and destroys the regular iambic pentameter ('While thundering terrors man's whole island shake'). The language of *Order and Disorder* becomes Lucretian in its violent recognition of physical instability. Referring to the 'pure innocence' of prelapsarian skin, 'their robe of glory and defence', Hutchinson informs us that

---

[27] Hutchinson translates from Lucretius:
> Againe when we
> Those mooving attoms in the sunbeames see,
> The *perplext agitations* there declare,
> Such *secret tumults* in the matter are. (italics mine)

See *The Works of Lucy Hutchinson, Vol. 1: The Translation of Lucretius*, ed. Reid Barbour, David Norbrook and Maria Zerbino (Oxford, 2012), Book 2, lines 122–25 (p. 91). For 'by impulsive force' see Book 2, line 62. All further quotations from Hutchinson's Lucretius are taken from the first volume of this edition and will be referenced in the text by book and line numbers.

'sin tore that mantle off' (4. 255–57). From this point onwards, mankind is pictured as physically and spiritually dissolute.

In her translation of the Lucretian report of atomic dissolution (Book 2 of *De Rerum Natura*), Hutchinson purposefully evokes a supernatural presence – the 'stronger blow / Then nature can resist' – to bring the Epicurean belief in contingent existence more comfortably in line with her own principles of divine preordination. When, upon writing *Order and Disorder*, she comes in turn to Christianise moments from her translation to form contributions to her biblical poem, direct allusions to the deity are often dropped to accentuate the threatening insecurity and isolation of postlapsarian corporeal life. This draws clear parallels with Scott-Baumann's reading of the Lucretian world 'in decay', as a world 'without the Christian God at its centre'. The 'stronger blow' behind the destruction in *Order and Disorder* is the withholding of divine influence – the consequence of physical transgression. God becomes distanced from the fallen world but, as Adam warns Eve, these are no random 'shafts' of cosmic violence (5. 518). Hutchinson reminds us that all acts of atomised dissolution serve a predestined purpose.

The parallels between Hutchinson's Lucretius and her accounts of physical and spiritual ruin in *Order and Disorder* are many. What has not been documented are the more spiritually productive, ultimately positive ways in which atomic language enters the Christian poem. The focus of this chapter is on how Hutchinson's atomic understanding of matter deepens the poet's – and the reader's – sense of empathy with the loss of prelapsarian bliss; of what it is to feel, at the level of corpuscular instability, the realities of the postlapsarian condition. This empathy, an appreciation of and effort to reconnect with first principles, not only results in pity at human loss but leads to the promise of future resurrection, a purified renewal of being that Hutchinson expresses in atomic and chemical terms. Jonathan Goldberg has touched on this in his account of Hutchinson's Lucretian 'seed', carried over into *Order and Disorder*: 'Grasping how the insensate becomes sensate, the sensate insensate, one grasps the principle of the seed, which is the basis for the fact that nothing remains itself for ever and that nothing ever vanishes materially.'[28] Atomised postlapsarian being in *Order and Disorder* follows this cycle of generation and corruption and links to the underlying constancy of God. Human experience of disorder and change unveils the promise of divine order at the first and last of things, a connection perceived initially in the poem by Hutchinson's Adam, the first principle of mankind.

In the first part of this chapter, I introduce the compelling link between Adam and the indivisible atom, both within early modern intellectual culture more broadly and between the pages of Hutchinson's creation poem. In the second section, I turn to examine in detail Hutchinson's chemical view of atomic regeneration in *Order and Disorder*. While these processes of transformation are based in corpuscular terms, it is through the chemical act of 'melting', both physically and in compassion, that the 'scattered atoms' (5. 253) of postlapsarian beings can be reconstructed by Christ. From a concoction of Lucretian poetics, imagery from the Geneva Bible and corpuscular chemistry, her account of Genesis introduces a glimmer of hope for the fallen as human beings 'melt' in the experience of empathy to be renewed in divine love. This is, therefore, more than a study of Lucretian resonances in Hutchinson's later work. *Order and Disorder* reveals a complex, spiritual poetics of the 'atom' which holds the key to reforming the '*World Made and Undone*'.

FIRST PRINCIPLES: ADAM AND THE ATOM

In hinting at the tantalising idea of a resonant homophone between 'Adam' and the 'atom' in seventeenth-century English culture, Joanna Picciotto has suggested a bond between two very different kinds of first principle, which could be at the heart of why Hutchinson drew so strongly on the atom to express her ideas about creation and resurrection. Picciotto writes, briefly yet suggestively:

---

[28] Jonathan Goldberg, 'Lucy Hutchinson Writing Matter', *ELH*, 73.1 (2006), 275–301 (290).

It is not far-fetched to suppose that the homophone of *Adam* and *atom* reinforced the link between historical and ontological 'originality'. The pun suggests that the search for the radical and primary cause of all experience terminates at once in the atom and in the original witness of creation who first discerned its qualities.[29]

The enjoyment of this pun in natural philosophical and theological writings remains subtle, but it is certainly not too far-fetched to claim that the parallels between these origins were crucial to how atomic thought provided a means for reasoning about matter and the divine – and moreover the ways in which Christian theology influenced, and even facilitated the acceptance of, contemporary interpretations of the atom. While many supporters of atomism were extremely wary of attributing the indivisible particle any agency – across seventeenth-century texts atoms were labelled 'senseless', jostling in random chaos or controlled by God – both the atom and the first man could be understood as end-points of creation at which it would be possible to understand all things.[30] Adam, the 'original witness of creation', saw all things in the natural world when they were in their state of perfection. Like Adam, the atom – an indivisible substance at least as old as the creation itself, within and integral to the form of all created things – could be understood as a seed of natural and spiritual potential: the simple truth behind all mysteries of the creation and the point from which

---

[29] Picciotto, pp. 199–200.

[30] For just three examples of this pervasive, disparaging phrase, see Seth Ward, the mathematician and bishop, speaking on 'Publique Judgments': 'They proceed not from a casual coincidence of various senseless Atoms'. *The case of Joram, a sermon preached before the House of Peers* (London, 1664), p. 7. Secondly, the jurist Matthew Hale, who remarks 'that there should be another kind of semem humanum or animale than what is moulded in the Bodies of Men or Animals, and elicited from them by a coincidence only of stupid, dead, and senseless Atoms, seems below the Genius of a Philosopher'. *The primitive origination of mankind, considered and examined according to the light of nature* (London, 1677), p. 260. Finally, the classicist and theologian Richard Bentley, who dismisses 'this sottish opinion of the Atheists; That dead and senseless Atoms can ever justle and knock one another into Life and Understanding.' *The folly and unreasonableness of atheism* (London, 1699), p. 56.

all physical entities came into being. The two subjects are united by a shared simplicity, a mutual proximity to the original divine cause.

As if to emphasise this status of Adam as the fount from which all future generations flowed, he was referred to widely in seventeenth-century literature as '*the* Adam', for reasons that were laid out helpfully by the Anabaptist Francis Bampfield: '*Adam* ... was the *Adam*, the common Head, and Representative of all Mankind, which proper name to make him more known, has a demonstrative notificative Particle before it'.[31] Bampfield's striking use of 'Particle' in this context carries the old grammatical sense of a small, indeclinable word, strengthening the ties further between the first man and the first principle of matter. As both proper and common noun, the Adam – the origin and end of all knowledge – came to adopt a variety of meanings parallel to the divergent interpretations of the 'atom'. The English Epicurean Walter Charleton generalised the noun 'Adam' to the extent that it came to signify the original state of any object. In his first major work, *The Darknes of Atheism*, he promotes philosophical enquiry with the claim that all chains of reasoning must eventually 'goe so high as the Adam, or Grandfather Idea'. Several years later, in *Physiologia Epicuro-Gassendo-Charletoniana*, he discusses the atomic cause of motion, referring to 'the inhaerent Gravity of ... Materials, Atoms' as the 'FIRST, that Adam or Radical and Primary Cause of all motion competent to Concretions'.[32] Charleton's amalgamation of the 'Adam' and the 'Atoms' implies that the pursuit of both 'Primary Cause[s]' would lead to the same discovery of how things came to be, what he referred to in *The Darknes of Atheism* as 'the Archtype

---

[31] Francis Bampfield, *All in one, all useful sciences and profitable arts in one book of Jehovah Aelohim* (1677), p. 138.

[32] Walter Charleton, *The darknes of atheism dispelled by the light of nature, a physico-theologicall treatise* (London, 1652), p. 10; *Physiologia Epicuro-Gassendo-Charltoniana, or, A fabrick of science natural, upon the hypothesis of atoms founded by Epicurus repaired [by] Petrus Gassendus; augmented [by] Walter Charleton* (London, 1654), p. 436. Joanna Picciotto notes Charleton's generalisation of the 'Adam' in his *Physiologia* in *Labors of Innocence*, p. 199.

or Protoplast, wherein all ... reality is inherent Formally'.[33] Adam was commonly dubbed the 'protoplast' – the first member, or model of the human race – in early modern culture; likewise, atoms represented the primary constituents of the natural world. Both first principles directed thought back to the creator that fused the primary materials together at the beginning of time. For Francis Bacon, the 'dignitie of knowledge' was in the 'Arch-tipe or first plat forme, which is in the attributes and acts of God'.[34]

Tentative yet tangible, the link between Adam and the concept of the indivisible particle was present in early modern culture. There is another reason for a strong connection between the two that extends beyond the shared characteristics of simplicity and longevity. For many seventeenth-century scholars, Adam was perfect to the extent that he possessed what would now be considered superhuman qualities. Not only was Adam himself *like* the atom, he had microscopic vision enhanced enough to perceive atomic structures; he was able to survey the heavens with a single glance, intuitively to recognise the nature of all things, and to experience fully what in the postlapsarian world could only take the form of hypotheses, or thought-experiments. According to Henry Vane, for example, he had an 'intuitive prospect into the nature of all visible and bodily things *in their causes*' (emphasis mine).[35] Interpreters took Genesis 2.19 as evidence for Adam's innate encyclopaedic knowledge and perfect linguistic ability:

> Now the Lord God had formed out of the ground all the wild animals and all the birds in the sky. He brought them to the man to see what he would name them; and whatever the man called each living creature, that was its name.

With perfect intuition, and comprehensive understanding, the first man was able to perform God's work in Paradise. By matching all living things with their proper, permanent names, he did more than

---

[33] Charleton, *The darknes of atheism*, p. 10.
[34] Francis Bacon, *The two bookes of Francis Bacon. Of the proficience and aduancement of learning, diuine and humane To the King* (London, 1605), p. 27.
[35] Henry Vane, *The retired mans meditations, or, The mysterie and power of godlines* (London, 1655), p. 53.

attach signatures – he completed the process of formation. The prelapsarian relationship between words and things was so perfect that to bestow names upon the creatures was identical to assigning them their natures. Consequently, Adam became somewhat of a mascot of ideal learning for seventeenth-century natural philosophers.[36] Francis Bacon's reading of Genesis led him to believe that learning – and moreover, learning how to learn – was Adam's first obligation in Paradise: 'the first Acts which man perfourmed in Paradise, consisted of the two summarie parts of knowledge, the view of Creatures, and the imposition of names'.[37] Many others followed suit in envisioning the first man as a figure of infinite capability, endowed with sensorial perfection and in complete harmony with the cosmos. Thomas Traherne intimates that he 'did Eat Herbs and Drank Water, but saw him self Advanced above the Skies: for the Sun and Moon and Stars … realy ministered to him'; his words nicely complement the visionary moment enjoyed by *Paradise Lost*'s Adam in Book 8, in which the giddy heights of the cosmos contract into a single act of prelapsarian enquiry.[38] For Samuel Pordage, author of a large-scale creation epic that preceded Milton's in print by six years, Adam was the perfect superhero, equally mobile by sea, land and air:

> He saw th'row all things, knew what all things meant:
> Gave names to all the Creatures, and did frame
> Them, as their natures so he gave their Name.
> Nor did he want the Camel, nor the Horse
> To carry him, he in himself had force
> Enough to move his Body, and to bear
> It where he list, o're Sea, or th'row the Ayr.
> No water could his Body drown, nor fire
> Consume; nor subject was't to Death's dread ire.[39]

---

[36] For more on this, see Philip Almond, *Adam and Eve in Seventeenth-Century Thought* (Cambridge, 1999).
[37] Bacon, *Aduauncement of learning*, pp. 28–29.
[38] *The Works of Thomas Traherne*, ed. Jan Ross, 9 vols (Cambridge, 2005–), II, *Commentaries of Heaven*, 'Adam', p. 215.
[39] Samuel Pordage, *Mundorum Explicatio* (London, 1661), p. 59.

If Adam had no need of a flying-machine, another popular opinion was that he had been blessed with such sensitive sensorial perception that moderns could not even come close with their microscopes and telescopes. The original man, Joseph Glanvill reminds his reader in *The Vanity of Dogmatizing*, had no need of 'Galilaeo's tube'.[40] Both Hutchinson and Milton draw on this cultural backdrop in their depictions. Eve as well as Adam shines forth 'Truth, wisdom, sanctitude severe and pure, / Severe, but in true filial freedom placed' in *Paradise Lost*, though it is Adam, designed for 'contemplation', who can hold intellectual conversation with angels (4. 293–97). Hutchinson is quick to assert that human intelligence 'cannot be / Refined enough for angels' company', most likely in critical response to Milton's epic, but her Adam is furnished with a 'comprehensive understanding', an ability to know all things with 'heaven and earth thus centred in his mind' (3. 297–98; 215; 231). Her perfect Adam, rather like the spiritual atom or monad in the tradition of Cusanus, forms a centre of contraction and understanding of 'All in All'.[41]

The problem of course is that subsequent, fallen generations of mankind lost the means to acquire this knowledge. What remains is the human body – a tangible reminder of what Adam was, what he became after sin, and what he would become. In a treatise on the possibility of 'total mortification of sin in this life', the Protestant divine William Parker revisits the familiar 'foure *Adams* mentioned both in the Scriptures and other writers', introducing each figure as follows: the 'first is our natural or earthly man'; the 'second is an inward portraiture of righteousnesse, a glimpse of the heavenly man, yet earthy, naturall and mutable; which we all beare at the first, and so are said *to have born the image of the earthly*, 1 *Cor.* 15.49'; the third is 'the old man of sin', and the fourth 'is the last Adam',

---

[40] Joseph Glanvill, *The vanity of dogmatizing, or, Confidence in opinions manifested in a discourse of the shortness and uncertainty of our knowledge, and its causes* (London, 1661), p. 5.

[41] Traherne uses the title 'All in All' as one of his topics in the *Commentaries* (III, p. 11).

Christ.[42] Parker breaks down the several ways in which mankind could relate to their first principle in the present day. Our bodies, if not as perfect or beautiful, were fashioned in the form of the first man. Simultaneously, these bodies remind us of our inherited sin, the 'old man', but ultimately, and most importantly, flesh promises salvation in the form of the incarnate Christ. The second Adam is an embodiment of this promise: the 'earthy, natural and mutable' mortal man still offers a glimpse of the 'heavenly'. Yet Parker guides us to Paul's words in 1 Corinthians 15.49: 'Just as we have borne the image of the man of dust, we will also bear the image of the man of heaven.' Intriguingly, and shedding light on one of the several ways in which 'the Adam' became associated with a special privileging of the atom, 1 Corinthians 15 continues with the following:

> Listen, I will tell you a mystery! We will not all die, but we will all be changed, in a moment, in the twinkling of an eye, at the last trumpet. For the trumpet will sound, and the dead will be raised imperishable, and we will be changed. For this perishable body must put on imperishability, and this mortal body must put on immortality. (1 Cor. 15.51–53)

Parker's reference to mutability, and the promise from Corinthians that the perishable body will rise imperishable 'in a moment, in the twinkling of an eye', introduce the suitability of the atom as an accommodative framework for thinking through the Day of Judgement. The 'twinkling of an eye', 'in ictu oculi' in the Vulgate, was acquainted in early modern culture with the smallest possible measured 'moment' – the 'Atom of Time'. From the early Middle Ages, the term 'atom', or 'atomus' after the Latin, was used to refer to the smallest unit of measured time: there are 376 atoms in a minute, or 22,560 in an hour.[43] This significance may well have been considered archaic by the sixteenth century, but associations between the single atom and the representation of time remained

---

[42] William Parker, *A revindication set forth by William Parker, in the behalfe of Dr. Drayton deceased, and himself of the possibility of a total mortification of sin in this life: and, of the saints perfect obedience to the law of God* (London, 1658), p. 7.

[43] See Andrew Pyle, *Atomism and its Critics: Democritus to Newton* (Bristol, 1997), p. 195.

common far into the seventeenth century.[44] While the Latin describes the miraculous transformation of 1 Corinthians 15 taking place 'in momento', the Greek refers to an instant, ἄτομος – or 'atomos'.[45] Consequently, a single 'atom' commonly represented the otherwise unfathomable period of time between death and resurrection.

Lexicographer Thomas Blount expounds this idea in 1654, mentioning in his study of rhetoric *The Academie of Eloquence* that 'Death is that inconsiderable atome of time that divides the body from the soul'.[46] The Independent minister Thomas Beverley was very fond of mentioning 'the Atom of Time' in his treatises. Like many dissenters of the period, Beverley drew heavily on Paul's letter to the Corinthians, and 1 Corinthians 15.51–53 especially, writing on one occasion of

> the *Saints*, The *Living*, the *Remaining*, who *dye not*, but are *Chang'd*, are *Chang'd* in the Atom of Time, the very same *Moment* in the very *Twinckle of an Eye* with the *Saints Raised*, 1 Cor. 15. 51, 52[47]

---

[44] To give just two examples: in a sermon on Ecclesiastes 12.1, the churchman and historian Thomas Fuller meditates on the message 'Remember *now* thy Creatour' with the explanation 'I say *now*, now is an Atome, it will puzzle the skill of an Angell to divide' (1655). The 'Atome' he has in mind is clearly the 'Atom of Time'. Thomas Fuller, *A collection of sermons* (London, 1655), p. 29. Ten years later, in John Crowne's prose romance *Pandion and Amphigenia* (1665), the kidnapped nun Glycera begs that

> when that atome of time, that inconsiderable instant that links past and future, shall unlink my soul from my body; then, oh then, let it arrive at that blest Elizium, where one eternal indivisible moment chains refined souls to an immortality of unconceivable felicity.

John Crowne, *Pandion and Amphigenia; or, The history of the Coy Lady of Thessalia Adorned with Sculptures* (London, 1665), pp. 203–04.

[45] For more on this see David A. Hedric Hirsh, 'Donne's Atomies and Anatomies: Deconstructed Bodies and the Resurrection of Atomic Theory', *Studies in English Literature 1500–1900*, 31.1 (1991), 69–94 (83).

[46] Thomas Blount, *The academie of eloquence* (London, 1654), p. 62. This metaphysical use of the word 'atom' entered the English language earlier in the century. See also, for example, George Rivers's *The Heroinae* (London, 1639), where an 'atom' is an 'inconsiderable' amount of time between death and resurrection (p. 33).

[47] Thomas Beverley, *An appeal most humble yet most earnestly by the coming of our Lord Jesus Christ, and our gathering together unto him, even adjuring*

On another occasion, in a sermon on Matthew 25.13, Beverley offers his reader some exposition on the recurring 'Atom of Time':

> What a little time is an *Atom* of Time; it is the smallest, most undividible for littleness, we can conceive; Now, saith the Apostle, 1 *Cor.* 15. 1, 2. [sic] *In an Atom or Moment of Time, in the Twinkling of an Eye, the Dead shall be raised incorruptible; and we the living Saints shall be changed:* None can express or think this swiftness.[48]

An atom, for Beverley, is the smallest, 'most undividible' portion of time. His claim that we can 'conceive' of its littleness, but 'none can express or think' its 'swiftness', presents an irresolvable paradox inherent to the atomic concept. *Atom* is, of course, a negative word; an atom can only be described by what it is not – indivisible, invisible; according to some beliefs, immortal. Unfathomable in many ways itself, the atom could form an appropriate object of accommodation for theological obscurities.

Parker does not reference atoms in his treatise on the mortification of sin, but the status of 1 Corinthians 15 as the source both for his exposition of the Adamic body and the 'Atom of Time' invites further parallels between the first man and the indivisible particle. Like Parker's Adamic human body, atoms associated with acts of resurrection also offer 'a glimpse of the heavenly'; an insight into the potential for matter to 'put on immortality'. The 'earthy and mutable' body bears the stamp of its potential for spiritual transformation through its physical inheritance of the first and last Adams: the archetype and the incarnate Christ. Similarly, atoms – the founding particles of all things, which form, unform, and re-form created entities – were associated with resurrection not only through the 'Atom of Time', but from their activity as the agents of generation and corruption within

---

the consideration of the most contrary minded who love his appearing concerning the Scripture on due compare, speaking expresly, or word for word (London, 1691), pp. 3–4.

[48] Thomas Beverley, *The parable of the ten virgins in its peculiar relation to the coming and glorious kingdom of our Lord Jesus Christ opened according to the analogy of the whole parable, and of Scripture in general, and practically applied for exercising all the churches to holy watchfulness* (London, 1697), p. 82.

all creatures. The association between the raising of bodies at the last trumpet and the recongregation of scattered particles was so prevalent in seventeenth-century literature that it is actually surprising that Parker makes no direct reference to dust or atoms.[49] Unsurprisingly, Hutchinson does as she turns to the fate of Adam and Eve.

OF MELTING, CONDENSATION AND COMPASSION

1 Corinthians is one of the books Hutchinson cites most regularly in *Order and Disorder*, and coincidentally something of a puritanical favourite (John Evelyn's diary complains routinely during the early years of the Interregnum that all the Sunday sermons were on Corinthians 1 and 2 and the 'mysterious' subject of predestination).[50] Following their corruption through Sin, Adam and Eve

---

[49] For example: 'I believe then, 1 Cor. 15.52. that at the shrill sound of the last Trumpet, which is no other, than the mighty, and powerfull voice of our Arch-Angel, Christ Jesus, the innumerable Atoms of my Body shall incorporate'. James Harrington, *Horæ consecratæ, or, Spiritual pastime* (London, 1682), p. 164. Also, John Scott writes in 1687: 'as by a loud voice or a Trumpet it was anciently the custom of the Jews, and other Nations, to summon Assemblies, and particularly by a Trumpet to collect and rally their Armies; so at the Resurrection our Saviour by the Ministry of his Angels, under the conduct of their Archangel, will assemble and rally our scattered Atoms'. *The Christian Life* (London, 1687), p. 507. Also in 1687, Benjamin Calamy writes: 'All the atoms and particles into which mens bodies are at last dissolved, however they seem to us yet are safely lodged by God's wise disposal in several receptacles and repositories till the day of restitution of all things ... till the sound of the last trumpet shall summon them, and recall them all to their former habitations.' *Sermons preached upon several occasions* (London, 1687), p. 384.

[50] Evelyn writes in his entry for 3 April, 1652: 'I went first to Church, the Minister preaching on 2. Cor: 5.17, somewhat mysteriously about predestination, & such high points, as the manner was: his drift was to shew in what sense the Elect were one with Christ.' He then comments in the following months how the minister continued to preach on 2 Corinthians, chapter 5 (concluding only on 15 January 1654). See *The Diary of John Evelyn*, ed. E. S. de Beer, 6 vols (Oxford, 1955), III, p. 82.

discover the promise that their earthly materials will re-congregate into renewed bodies, perfected by divine resurrection. As William Parker writes, we – the succeeding fallen generations – learn of what the first parents received first hand through the representative figure of 'the Adam'. In Hutchinson's poem the lines of 1 Corinthians, 15.51–53 are not far away as we make the discovery through the aid of corpuscular physics:

> When men out of the troubled air depart,
> And to their first material dust revert,
> The utmost power that death or woe can have
> Is but to shut us prisoners in the grave,
> Bruising the flesh, that heel whereon we tread,
> But we shall trample on the serpent's head.
> Our scattered atoms shall again condense,
> And be again inspired with living sense;
> Captivity shall then a captive be,
> Death shall be swallowed up in victory,
> And God shall man to Paradise restore,
> Where the foul tempter shall seduce no more. (5. 247–58)

In her marginalia, Hutchinson cites 1 Corinthians 15.20–22, 26, 54, 55 and 57 next to the final two lines of the above-quoted passage, referring always to the Geneva translation popular with puritanical readers. Although unreferenced, 15.51–53 clearly has a part in 'Our scattered atoms shall again condense, / And be again inspired with living sense'. Through the combined mediums of scripture, the Adam, and the atom, we learn mankind shall be restored once more to the original state of 'Paradise'.

While in *Paradise Lost* this promise might have been declared by Milton's rather verbose Adam, speaking directly of the prophetical knowledge he gained by the fruit, in *Order and Disorder* the foretelling of resurrection and eternal salvation is bequeathed – curiously – in the words of the narrator herself. Hutchinson claimed in her preface to the reader that her aim was only to discuss what could be found in scripture, 'that revelation God gives of himself and his operations in his Word'.[51] This explains her decision not to

---

[51] 'The Preface', p. 3.

attribute the above revelation to Adam and Eve, as she goes on to admit, regretfully: 'How far our parents, whose sad eyes were fixed / On woe and terror, saw the mercy mixed / We can but make a wild uncertain guess' (5. 259–61). This does not, however, explain her reasons for mixing citations from Old and New Testament sources with a distinctly atomic and chemical vocabulary, a concoction of discourses recollective of Hester Pulter's experiments in devotional verse.[52] As Paul Hammond has commented in reference to lines 253–56, the 'first couplet is pure Lucretius, the second pure St Paul.'[53] Even Hutchinson's reference to 'troubled air' is an imaginative amendment to the cited inspiration behind the line, Job 3.17–19, which reflects on those who, after death, as 'prisoners rest together, and hear not the voice of the oppressor'. The additional phrase 'troubled air' recalls the atomic violence of 'the air's quick-piercing shafts' that 'through / Both leaves and pores into the bowels flew' (4. 253–54).[54] The central, most powerful part of the promise – that 'Our scattered atoms shall again condense, / And be again inspired with living sense' – not only describes the resurrection of the body by the regeneration of its atomic particles, but claims that its materials will 'condense'. Hutchinson's wording brings a chemical and, thereby, spiritual dimension to the physical activity. A verb of alchemical origin, the vow that our dispersed atoms will 'condense' once more into a single form explains what will happen, at a corpuscular level, to the body following the Last Judgement. Both the detail on condensation and the concentrated focus on physical indivisibles emphasise the significant reconstruction of the human body, as it is converted into a vessel for the divine inspiration of 'living sense'. Hutchinson's natural philosophical

---

[52] For a study of Hester Pulter's atomic poetry, see Chapter Three.

[53] Paul Hammond, 'Dryden, Milton, Lucretius', *The Seventeenth Century* 16.1 (2001), 158–76 (166). See also Julie Crawford, who claims the first half of this couplet illustrates the 'robust materialism of Lucretian atomism'. Crawford, 88.

[54] The Geneva Bible only declares at Job 3.17, referring to the grave: 'The wicked have there ceased from their tyranny, and there they that labored valiantly, are at rest.'

resurrection brings a heightened physicality to her specified source for line 254, from Isaiah (26.19):

> Thy dead men shall live, together with my dead body shall they arise. Awake and sing, ye that dwell in dust: for thy dew is as the dew of herbs, and the earth shall cast out the dead.

Strikingly, Isaiah's 'dew of herbs' appears to have led Hutchinson to the alchemical practice of palingenesis, the idea that plants could be restored from their first principles.[55] The scripture suggests the parallel between the rebirth of the body and the sprouting of herbs through simile, but the strength of Hutchinson's proposal – our atoms '*shall* again condense' – transforms a biblical image into a viable natural philosophical theory.

Though the inevitability that men will 'revert' to their 'first material dust' is at first glance a biblical commonplace, Hutchinson's qualification that the dust is 'material' is suggestive. The line can be read two ways: either mankind returns to its first material, dust, or the sense is that '*material* dust' acts as a qualification of the term, implying that there is additionally something extra-material to the minute constituents of the human form. It tantalisingly suggests the possibility of an alternative understanding of 'dust' as a supernatural, and possibly immortal substance. In his biblical history *Volatiles from the History of Adam and Eve* (1674), John Pettus reads great significance into the precise information that mankind was formed 'by the dust *of the ground*' (italics mine), arguing that it was in part due to this lighter, inspired material that humanity came to be infused with spirit:

> by the dust of the ground is to be understood the superficial part of the earth, of which man onely was made; for all Creatures are said to be made out of the ground, but not of the dust of the ground[.]

He proceeds to expound the special properties of this dust:

> Now whither this dust was made by a peculiar omnipotent Calcination, or other Rarification is not demonstrated; but we may conceive it of the most attenuated part of the earth, and therein the

---

[55] I discuss Hester Pulter's poetic interest in palingenesis in Chapter Three of this book, pp. 130–32.

more Noble part, because capable of the most activity ... [I]f the ordinary dust produceth such agile Creatures [as other animals], we may collect that our Creator hath adapted to our more agaile body (in relation to our Souls) far more agile dust then to other Creatures. For the Targum of Jerusalem adds to our honor, that we are made *ex palvere Sanctuarii*, i.e. of holy dust, differing from all other dusts: which should raise this Contemplation in us, that as we are not like beasts or other Creatures in our Temperaments, they made of the ground, that is, of the faeces or dregs of the Earth, we of the Superficies (or of some peculiar sanctify'd dust;) so ought the habits of our bodies to be sublime, and alwaies ascending to an higher sphere[.][56]

Pettus's vocabulary suggests, rather like Hutchinson's but more explicitly, that we might think of the divine creator as chemist. He finds a sure sign of mankind's superiority to the other creatures in the mysterious, spiritual properties of the *material* creation of our species. The dust, he posits, may have been the outcome of a mystical alchemical act by the divine ('Calcination'; 'Rarification') and, regardless of its origin, it stands that the dust forming our bodies is more refined than 'ordinary' earth.[57] Our bodies are more agile than those of the most agile creatures, he argues, because our bodies accompany an immortal soul; as Pettus continues, he gives the impression that this special dust, 'differing from all other dusts', not only hosts but is part of our immortal essence. This 'holy', 'agile' and 'sanctify'd dust' enables mankind to draw near the divine even while in the state of mortal embodiment – because of what we are made, he writes, 'so ought the habits of our bodies be sublime' – and promises to be refined further into spiritual purity following rebirth and resurrection. If 'common dust', argues Pettus, 'flies like Atoms over the Surface of the earth', who knows into what heights our indivisibles will take us?

---

[56] John Pettus, *Volatiles from the History of Adam and Eve* (London, 1674), pp. 21–23.

[57] Pettus's 'attenuated', rather like Hutchinson's 'condense', is a word that bore emphatic natural philosophical connotations in the seventeenth century. Referring to something rarefied, or thin in consistency, Francis Bacon and Henry More are among those listed as its first citers in the *OED*, 'attenuate, adj.' (1626 and 1647).

This idealisation of matter that is 'sanctify'd', to borrow the word from Pettus, is prevalent throughout Hutchinson's descriptions of created things in their rightful, or rejuvenated order. On this note, acts of atomic and chemical transformation move into a different remit in her reflections on soteriology. As her focus shifts closer to aspects of redemption, her reflections on transformation become less blatantly atomic and more chemical. The reason for this is that, while her atoms and dust are associated more with the formation (and deconstruction) of divine bodies, Hutchinson chooses to extend her alchemical vocabulary even to the state of the immortal soul, frightened by sin:

> Mercy still doth fainting souls revive,
> And in its kind embraces keep alive
> A gentler fire than what it lately felt
> Under the sense of wrath. The soul doth melt
> Like precious ore, which when men would refine
> Doth in its liquefaction brightly shine;
> In cleansing penitential meltings so
> Foul sinners once again illustrious grow,
> When Christ's all-heating, softening spirit hath
> Their furnace been, and his pure blood their bath. (4. 345–54)

Once again, God emerges as the preeminent alchemist. The divine melts the soul as though it were a metallic 'ore', both 'precious' and profitable. Hutchinson follows the imagery of medieval alchemical literature in her reference to the 'bath', an image that was commonly used to represent the vessel bearing the chemical experiment.[58] This suggests a literary knowledge of Christianised chemistry that joins the alchemical process in metaphor with Christ's purging of sin, an association that was not absent from early modern devotional poetry (the Catholic Robert Southwell, for example, imagines the 'Burning Babe' Jesus melting 'into a bath, / To wash them [sinners] in my blood'.)[59] Hutchinson cites Revelations 1.5 in the margin next

---

[58] See Lyndy Abraham, *A Dictionary of Alchemical Imagery* (Cambridge, 2001), pp. 17–18, for an introduction to the image of the chemical bath.

[59] Robert Southwell, 'The Burning Babe', quoted from *The Complete Works of R. Southwell, S. J., with Life and Death* (London, 1876), pp. 98–99.

to this passage, one of several scriptural instances that affirm the power of Christ's blood to cleanse humanity from sin. The Geneva Bible specifically refers to the act of washing. It evokes the revelation of him 'that loved us, and washed us from our sins in his blood', therefore supporting the possibility of an alchemical reading of Christ's passion.

It is not, however, the incarnate Christ who is melting in Hutchinson's poetry, or even the fallen human body, but the sinner's *soul*. This is surprising for two reasons. Firstly, while bodies melt frequently across early modern poetic spaces – from overheated lovers burning in ardour, to dissolving corporeal prisons permitting spiritual escape – souls necessarily remain intact as indivisible and immortal centres. Secondly, where melting occurs in a biblical context, as it does frequently across both the Geneva and King James translations, it is typically a violent and fearful act from a displeased God. As Andrea Brady has noted, the verb 'to melt' is 'used twelve times to refer to the obliteration of the universe by the wrath of God, and seven times to the obliteration of people'.[60] Hearts 'melt' in fear at the prospect of battle or at the terrible power of an angry divine; the Psalmist sings of mountains at risk of dissolving away; the displeased deity of Ezekiel 22. 20 threatens to melt 'the house of Israel' just as 'they [people] gather silver and brass, and iron, and lead, and tin into the midst of the furnace, to blow the fire upon it to melt it' (GNV). The vocabulary in the example from

---

Katherine Eggert has offered the following analysis of Southwell's penitential alchemy:

> This vision owes much to nonalchemical sources, including the Petrarchan lover's suffering simultaneous cold and heat and Ignatius Loyola's meditations on the transcendent warmth of the Savior's birth in the cold of winter. But the fact that the fire in the incendiary babe's 'faultless breast' is fueled in a 'furnace' means that the vision also evokes an ideal alchemy, one that successfully purifies 'The mettall in this furnace wrought, / ... mens defiled soules'.

Eggert, *Disknowledge: Literature, Alchemy and the End of Humanism in Renaissance England* (Philadelphia, 2015), p. 55.

[60] Andrea Brady, 'The Physics of Melting in Early Modern Love Poetry', *Ceræ: An Australasian Journal of Medieval and Early Modern Studies*, 1 (2014), 22–52 (32).

Ezekiel offers parallels with Hutchinson's alchemical reading of spiritual purification, but this destruction is not like the 'cleansing penitential meltings' (4. 351) promised by *Order and Disorder*. There are a few occasions, found only in the Geneva translation and not in the King James Version, where melting is associated with the act of repentance and receives a merciful, rather than wrathful, response. The God of 2 Kings 22. 19 delivers a message in this vein to the King of Judah, informing him that 'because thine heart did melt, and thou hast humbled thyself before the Lord when thou heardest what I spake against this place ... and wept before me, I have also heard it'. A parallel passage to this appears in 2 Chronicles 34. 27. With a similar sentiment, the singer of the Psalms exclaims 'My soul melteth for heaviness: raise me up according to thy word' (Psalm 119. 28). In these cases, the heart and soul melt not so much for fear but for shame and remorse. Melting is the act of prostrating oneself in front of the Lord; the ultimate, internalised motion of Christian humility, accompanied by external signs of penitence (weeping and tearing clothes.)

It is likely that Hutchinson had these penitential episodes in mind when composing the passage about mercy from *Order and Disorder*. The tone is nevertheless dramatically different from any of her biblical sources and contrasts with passages elsewhere in her creation poem, which focus in frightening detail on the dissolution of the sinful world, its 'tender body rent' by 'whirling fires' (5. 326). The flame of mercy, on the contrary, is a 'gentler fire', which revives and maintains 'fainting souls'. As a chemical process, the act of melting is figured here as gentle and nurturing; the emphasis is on the promise of generation over corruption. To melt is to lose substantiality of individual form, but not in the negative sense of the body assailed by warring elements and winds, as it is in the immediate aftermath of the Fall. Melting is also an empathetic breaking-down of the barrier between self and other, an act of love and of letting the other in. Hutchinson's extension of this to the soul heightens the intimacy of exchange between human being and Christ. Joined in the 'bath', the blood of Christ and the fainting soul recall the Red King and White Queen of alchemical symbolism, respectively the active principle and the passive, indefinite form, melted together by Christ's 'all-heating softening spirit'.[61] The reference to the 'bath' and the three subjects – active blood, passive soul and connecting spirit – suggests that Hutchinson was familiar with alchemical allegory and deliberately drawing on its significance. She may have had in mind the images of the *Rosarium Philosophorum* (1550), which depicted the spirit between the Red King (sulphur) and the White Queen (mercury) as a dove, an immediately recognisable symbol for the Holy Spirit.[62]

The result of the union is a perfect, reformed substance, a soul that is once more 'illustrious' (an adjective that bore the significance during the seventeenth century of something 'lit up', in addition to its endured sense of eminence). The soul is both re-formed and reformed in Christ. This promise builds on the prophecy of atomic recongregation explored in the following canto, the vow that 'Our scattered atoms shall again condense, / And be again inspired with living sense'. In the earlier passage, it is not the body but the immortal soul that 'shall again condense' – and not without undergoing a vital transformation of its properties. Hutchinson extends alchemical–atomic language to the repentant soul to stress its endurance as a first principle, like the atom. There is a strong connection here with corpuscular alchemy, an association in Hutchinson's work further supported by the significant number of alchemical terms, including the verbs 'condense' and 'rarefie', she included in her translation of Lucretius.[63] Within chemistry, as Daniel Sennert wrote in 1619, observing the dissolution of a substance and its consequent reappearance proves the existence of atoms. Silver and gold may be melted down with aqua fortis so that 'no metal can be detected in the water by sight', but since they are 'really present', the metals may reappear in their own pure and segregated forms – such a phenomenon must occur because the atoms in the silver and gold

---

[61] See Abraham, *Dictionary*, pp. 167–68, for further information on the symbolism of the 'red man and white woman'. See also in Abraham 'king' (pp. 110–13) and 'queen' (pp. 161–62).
[62] *Rosarium Philosophorum* (Frankfurt, 1550).
[63] See Book 1, 626–60, especially: 'The lower fires to litle purpose were Condenst or rarefied, if all parts here / The nature of the element possest'.

recongregate and rebuild the original metals.[64] Hutchinson never conceives of the rational soul literally as formed by atoms: to do so would be the height of Epicurean atheism, as she herself proclaims in the prefaces and marginalia to her works. This did not prevent, however, the vocabulary of atomic–alchemical transformation lending itself to the conclusions she sought to draw about dissolution, resurrection and preordination. The 'liquefaction' of 'precious ore' provided a powerful analogy for Christ's purification of the soul, a metaphor supported by numerous references in the Bible to 'melting' but pushed much further by her knowledge of atoms and alchemy. It is specifically Hutchinson's *corpuscular* reconfiguration of what it is to 'melt' in *Order and Disorder* that inspires hope and positivity. She outwardly rejected atomism as an idea received from classical paganism, but envisioning substances broken down into indivisibles made possible the distillation and regeneration of postlapsarian life. The endurance of first principles, figured in chemical and atomic terms, promised rebirth at the point of origin.

For Hutchinson in *Order and Disorder*, episodes of the chemical return to the soul's purified origins occur at moments of deep empathy, with recognition of human sin and the divine promise while in the midst of melting, both physically and emotionally. It is the first principle of mankind, Adam, who understands the importance of empathetic connection at the level of shifting particles and forms. Following the Fall and God's subsequent judgement, Adam recognises the future redemption that lies in faith. He listens to Eve's plaint, in which she questions why 'should we not our angry Maker pray / At once to take our wretched lives away?' (5. 411–12) Adam is moved, responding with a 'melting soul' and the reassurance that God will be there to soften their extremes. God, Adam reminds Eve:

> Nor only sees, but darts on us his beams,
> Ministering comfort in our worst extremes.
> When lightnings fly, dire storm and thunder roars,
> He guides the shafts, the serene calm restores. (5. 515–18)

---

[64] Daniel Sennert, *De Chymicorum* (1619), p. 362. Translated and quoted by William R. Newman, *Atoms and Alchemy: Chymistry and the Experimental Origins of the Scientific Revolution* (Chicago, 2006), p. 99.

Hutchinson's Adam bypasses his equivalent in Milton's epic, who needs the archangel Michael to remind him of God's omnipresence in his despair at leaving Eden. The Adam of *Paradise Lost* is 'Heart-strook with chilling gripe of sorrow' at the thought of leaving Paradise (11. 264), his heart frozen in cold and crippled by emotion – the opposite to Adam's 'melting heart' in *Order and Disorder*, which is warm and open in feeling. No angelic visitor nor external revelation of what is to come is required for Hutchinson's first man, who understands God's universal influence immediately and immanently. In communicating this, Adam directly counters Lucretius with his message that God 'guides the shafts'; that all motions, no matter how 'extreme', are controlled by the deity, predestined to conclude in 'serene calm'. The state here is opposed to Lucretius's violent and indifferent 'congressions' of matter, the language of which Hutchinson adopts in her poem during episodes of great upheaval and disorder. Instead, her Adam commands:

> Let us lie close in Mercy's sweet embrace,
> Which when it us ashamed and naked found,
> In the soft arms of melting pity bound,
> Eternal glorious triumphs did prepare,
> Armed us with clothes against the wounding air,
> By expiating sacrifices taught
> How new life shall by death to light be brought. (5. 560–66)

Using Adam as mouthpiece, Hutchinson rewords her promise of the soul's chemical regeneration from the end of Canto 4 ('Mercy still doth fainting souls revive, / And in its kind embraces keep alive / A gentler fire … / … The soul doth melt.') While the prior language of Hutchinson/the narrator is more explicitly alchemical, Adam nevertheless seeks the embrace of Mercy's 'soft arms of melting pity', in recognition that 'new life shall by death to light be brought'. Adam's reference to the 'wounding air', like the 'troubled air' of some lines later, echoes the infliction at the moment of the Fall, when Adam and Eve's 'coverings were too slight and thin' to keep out 'the keen air's quick-piercing shafts'. The clothes provided by God offer better protection, as the newly 'armed' sinners are offered a typological reminder of their future salvation in Christ.[65] There is a slight ambiguity around the 'soft arms of melting pity', which belong on the one hand to personified Mercy but could also depict Adam and Eve 'bound' together in their shame and nakedness, with mutual compassion. Mercy teaches the first parents of mankind how to 'melt' and feel, through the chemical dissolution of boundaries between self and other. Hutchinson's narrator in turn inherits this from her shared experience with the primary ancestors. Where she goes off-book in *Order and Disorder*, it is through familiarity and empathy with postlapsarian grief and atonement: 'Methinks I hear sad Eve', she writes in Canto 5, imagining the first mother bewailing 'her woeful state, with such sad plaints' (5. 399–400). It is through this intimate, inherited understanding of the fallen state and through knowledge, after Christ, of the future redemption that she can transfer her words (on the promised, future purification of the soul) to her Adam, the atom of mankind, merely a few lines later.

This understanding is not shared with his counterpart in *Paradise Lost*, who instead wanders in error through a soliloquy that expands upon the despair voiced by Hutchinson's Eve. It transpires that, in terms of an ability to read and understand the ways of the material cosmos, the first couple of *Order and Disorder* have the upper hand. While Hutchinson's retelling of the Fall can conclude with the promise that 'our scattered atoms shall again condense', Milton's Adam – rather like his Satan – seems to have focused more on the Epicurean attribute of 'senseless' matter than on its regenerative, chemical properties. Before and after the Fall, he continues to misunderstand the conditions of his creation, not appreciating – as the theologian John Pettus did – that human beings were made of a special kind of dust. In Book 8 he recounts his first day as a living thing, remembering how, after tiring himself out in learning to walk and run

---

[65] Hutchinson explains earlier in Canto 5, when introducing Adam and Eve's divinely wrought clothes: 'The skins of the slain beasts God vestures made / Wherein the naked sinners were arrayed, / Not without mystery, which typified / That righteousness that doth our foul shame hide' (5. 277–80).

> On a green shady bank profuse of flowers
> Pensive I sat me down; there gentle sleep
> First found me, and with soft oppression seized
> My drowsèd sense, untroubled, though I thought
> I then was passing to my former state
> Insensible, and forthwith to dissolve (8. 286–91)

This 'thought' was of course incorrect. Far from 'insensible', Adam dreamed and awoke to find his dream was truth; that he had been transplanted into the garden in which he was ordained to reign over all other creatures. At this point, Adam has not yet received the divine guidance that granted him the intuitive recognition of all things and the ability to assign names. The 'insensible' dissolution he imagines, not to mention his 'untroubled' state of mind in accepting this supposed fate, is textbook, if clichéd, Epicureanism. But his misunderstanding persists, even after his taste of prelapsarian perfection, and even after his taste of the Tree of Knowledge. In Book 10, Adam fearfully questions the meaning of 'death', and exclaims:

> how gladly would I meet
> Mortality my sentence, and be earth
> Insensible, how glad would lay me down
> As in my mother's lap! (10. 775–78)

While his contemplation of returning to dust is not as 'untroubled' as it was in his first day of innocence, Adam continues to misperceive the earth as 'Insensible', when of course the soil in the world of *Paradise Lost* is anything but. His wishful desire for the immediate stroke of death, combined with his reasoning that it is 'but breath / Of Life' responsible for sin, directs him to the heresy of mortalism – a belief asserted by Milton in *De Doctrina Christiana*, but here exposed as a misunderstanding prone to result in tragedy.[66]

In *Order and Disorder*, it is Eve rather than Adam who voices her doubts about what must come next. Hutchinson's first man, in

---

[66] See William Kerrigan's evaluation of Milton's mortalism in 'The Heretical Milton: From Assumption to Mortalism', *English Literary Renaissance* (1975), 125–66.

contrast to his equivalent in *Paradise Lost*, recognises the nature of their punishment and understands the need to trust in God:

> since our chief and immaterial part,
> Not framed of dust, doth not to dust revert,
> Its death not an annihilation is,
> But to be cut off from its supreme bliss (5. 469–72)

This passage occurs as Hutchinson grants her Adam and Eve a rare moment of independent opinion. Adam applies natural philosophical logic to reason that there is a part of their beings which, not subject to dust, must therefore have immaterial and thence immortal properties. Arguably, however, Hutchinson's Adam, though he understands the eternal order of soul and God, likewise fails to appreciate in full the special nature of divine 'dust'. On the basis of Hutchinson's exploration of their regenerative properties ('inspired with living sense', 5. 254), she would agree with John Pettus that the condensing atoms of the human being are no lifeless particles but the quasi-physical, quasi-spiritual entities that distinguish mankind from other natural creations.[67] Unlike the dull atoms of baser beings, the 'dust' composing Adam and Eve bears the supernatural capability for universal connection and resurrection. This resurrection takes place simultaneously, and paradoxically, in the divine renewal of individual selves and in the birth of future generations, all of whom carry 'the Adam' amongst their atoms.

---

[67] The vitality of Hutchinson's earth has also been noted by Julie Crawford, who observes 'an enspirited and respiratory sense of mobile materiality' in the natural world of *Order and Disorder*. Interestingly, it is in close relation to human beings that her nature assumes its vital characteristics, as humans' 'panting bosoms' exist 'in the same ecology as the "bosom" of the earth and "th'eternal bosom" of the waters ... as well as "th'air's soft breath" (3.114)'. Crawford, 81–82.

CONCLUSION: ATOMS AND EMPATHY

In both *Order and Disorder* and *Paradise Lost*, the presence of atomic movement exposes the relative harmony or disharmony between God and created individuals, whether positioned as the innate reality of the postlapsarian world or as unformed, potential matter on the margins of creation. This is especially the case in Hutchinson's poem, where allusions to atomic structure become exploratory descriptions, or assessments of spiritual health in the physical world. She drew on her advanced knowledge of Lucretius not only negatively – in intense corporeal descriptions of what happens with the abandonment of divine grace – but positively, to promote her quest for stable, ordered being.

The point at which atomic movement is made known to human experience in Hutchinson's poem, the Fall, is the point at which access to the microworld is withheld in Milton's epic. This is because the atom in *Order and Disorder* is not a vantage point of prelapsarian, natural philosophical privilege, but a broken-down particle made known at the onset of physical and spiritual distress. Despite its fraught, destructive origins, its accessibility is to the great advantage of postlapsarian beings: Hutchinson's atoms introduce the promise of resurrection in the divine, a rebirth imagined in atomic and chemical terms. Through the figure of Adam, the fallen first principle (or 'atom') of mankind, the narrator of *Order and Disorder* recognises that our 'scattered atoms shall again condense' and recongregate into spiritual perfection, an image that is rounded off in the poetry by references to the alchemical bath. Hutchinson's God takes form as the ultimate corpuscular alchemist, dissolving not only bodies but souls so to perfect their restoration in divine bliss. *Order and Disorder* merges a poetics of Lucretian atomism with the language of Christian mysticism, without acknowledging any ideological tensions as a consequence.

Despite, therefore, Hutchinson's own protestations in her Preface and disavowals from those who read her, the poetics of the atom in *Order and Disorder* was not reserved solely for negative, sinful situations. While atomic dissolution was associated closely with the Fall, the need to pick up the pieces of matter in reformation/

re-formation brought atoms and soteriology within the same remit. There is moreover an intimate connection between atomic reconfiguration and human empathy in the poem, as revealed by Hutchinson's focus on penitential 'melting'. Following the Fall, the barriers separating and protecting internal from external experiences dissolve, as atomic and chemical movement forms new connections between porous, grieving beings. Hutchinson's take on atomic being and soteriology through feeling encourages a rethinking of one of *De Rerum Natura*'s most famous, and severe moments – the pleasure that comes from witnessing others' misfortune, an episode in a materialist poem that would seem to counter the spiritual compassion promoted in *Order and Disorder*. Lucretius writes, in the words of Hutchinson's translation:

> Pleasant it is, when rough winds seas deforme,
> On shore to see men labour in the storme;
> Not that our pleasure springs from their distresse,
> But from the safetie we our selues possesse. (2. 1–4)

In her translation Hutchinson stresses that the pleasure we, the observers, experience does not stem 'from their distresse', but from recognition of 'the safetie we our selues possesse'. Her wording proposes some subtle changes from the original Latin. The pronoun is 'we', rather than the second person of Lucretius's text; as Barbour and Norbrook have argued in their recent edition of Hutchinson's *De Rerum Natura*, this lends the passage 'a proverbial effect' (*Works*, I, p. 531). This, combined with the added emphasis on 'safetie' and her substitution of 'Pleasant' for Lucretius's more sensuous '*suave*', creates a moment of feeling with the potential for transition into a Christian context. The chemical, atomic after-effects of the Fall in *Order and Disorder* reveal that it is necessary to experience disorder and change so to recognise, and be thankful for, order and constancy in God. The Lucretian act of witnessing the shipwreck offers parallels with the 'pleasure' that might spring from reading Hutchinson's creation poem, a gladness that emerges from observing the plight of postlapsarian being with, consequently, the knowledge that succeeding generations possess 'safetie' in divine mercy. The difference is that Adam and Eve, unlike the unfortunate sailors of *De Rerum Natura*, are not external others to the poet and readers, but essential parts of their being. It is at the exposure to disorder, through worldly suffering and reflection on the loss of the 'first Adam', that Hutchinson connects human experience with the essential movements of a corpuscular microworld, the indivisibles at the heart of creation that permit change, purification and the return to God.

It turns out that there is more to the link between 'Adam' and the 'atom' in Hutchinson and Milton's compositions than simple resonances with a wider cultural association. Adam, the atom of mankind, is required to come to know himself as a first principle, to fathom the divinely bestowed atoms of his being, so to reconnect with God and to reunite matter and spirit. The succeeding generations of humanity are made of the same atomic stuff. For these past, current and future believers, Adam is the original particle they seek: the archetypal figure, still spiritually and indeed physically present within the very components of their being, in closest contact with the deity. It is no coincidence that atoms have a place in the universes of each of the significant creation poems of the late seventeenth and early eighteenth centuries, from *Paradise Lost* and *Order and Disorder*, to Samuel Pordage's *Mundorum Explicatio* (1661) and Richard Blackmore's *Creation* (1712).[68] The atoms of matter and of text form a clasp between mortal, embodied experiences and the divine. In the attempt to trace a history of the untraceable, the very origins of time, atoms provide the means to reconnect with what has been lost. As Teilhard de Chardin would put it centuries later:

> To push anything back into the past is to reduce it to its simplest elements. Traced as far as possible in the directions of their origins, the last fibres of the human aggregate are lost to view and are merged in our eyes with the very stuff of the universe.[69]

---

[68] Richard Blackmore, *Creation: A philosophical poem demonstrating the existence and providence of a God in seven books* (London, 1712).

[69] Teilhard de Chardin, *The Phenomenon of Man*, trans. Bernard Wall (New York, 1959), p. 39.

Far from any fears of atheism or heretical Epicureanism, this 'very stuff of the universe' is that which comes closest to God. These 'scattered atoms' shall condense, Hutchinson's poem promises, not only into a resurrected human form but into renewed harmony between body and divine spirit.

AFTERWORD:
A POETICS OF THE ATOM

In his Cold War science fiction novel *Riddley Walker*, Russell Hoban reimagines the landscape of Kent two thousand years after nuclear war. Hoban writes in an envisioned future dialect: its sounds and spellings gesture at the interactions between a long-forgotten Christianity and terminologies of science and technology, traces of which remain only in the inhabitants' mythology. The myth that civilisation fell after the death of the 'Littl Shynin Man the Addom' who 'runs in the wud' conflates the splitting of the atom with the Fall of Adam.[1] What is left in the aftermath is a devastated physical landscape, but it is still 'humming' with spiritual significance. The concept of the atom has once again entered the remit of theology, as the protagonist Riddley discovers when he reaches the ruins of Canterbury Cathedral. He sinks to his knees with wonder, overcome with the realisation that the moving 'party cools of stoan' inspire 'the girt dants of the every thing' that 'keaps the stilness going'.[2]

Writing in 1980 and projecting his characters thousands of years into a postapocalyptic future, Hoban centres on a mystical connection between 'Adam' and 'the atom' very similar to the resonance between founding particle and human archetype in

---

[1] Russell Hoban, *Riddley Walker* (London, 1980), p. 30.
[2] Ibid., p. 160 and p. 163. Riddley's 'party cools' are the particles of matter, and the 'girt dants' suggests the dance of atomic motion.

seventeenth-century culture. Riddley's revelation that 'the girt dants', or great dance, of all things 'keaps the stilness going' is only a few steps away from Traherne's wonder that intrepid atoms, 'Without Change in themselvs, they all things change'.[3] The idea of reflecting on atomic movement to reconnect with a more perfect state of existence, once lived in the deep past but now lost, echoes Hutchinson's contemplations of Adam and atoms to rediscover the first principles of being. If the endurance of Hoban's imagery carries a degree of prophecy, the seventeenth-century poetics of the atom could have a long afterlife.

The poetics of the atom I argue for in this book was, in part, contingent and bound to the specifics of mid seventeenth-century intellectual culture. As the reception of atomic philosophy spread and interpretations of the word 'atom' expanded, understandings of the indivisible particle outgrew their natural philosophical origins. Meditations on the concept of an atom created new opportunities for poetical explorations of self, society and divinity. The characteristics of this poetics were several. Firstly, it was through its many appearances in poetry that the atom amassed further significance in seventeenth-century culture. Secondly, poets recognised the sympathies between atomic motion and the workings of poetic forms, which likewise aspired to create worlds and identities. This mobility was mirrored by the forms adopted by the authors of this study, especially the 'atoms' of individual verses or stanzas that were brought into generative, experimental sequences by More, Pulter and Cavendish. Authors like Traherne, on the other hand, concentrated on the potential of the singular atom as a connecting point between material existence and the divine. For Traherne in particular, it was the affinity between the volatile particle and the active, insightful movements of his lyric poetry that inspired his development of the atom as a reliable model for the similarly indivisible soul.

Thirdly, this book has revealed productive parallels between the potential of the indivisible atom and the creative capacity of poetic space. The atom, simultaneously the first and last of all things, offered in focus both a point of contraction and the promise of expansive,

[3] Thomas Traherne, *Commentaries of Heaven*, in *The Works of Thomas Traherne*, ed. Jan Ross, 9 vols (Cambridge, 2005–), III (2007), p. 358.

macrocosmic imaginative enquiry, limited only by the boundaries of creation. Likewise, the writers featured in this book understood the ability of poetry to conceive of an otherwise unfathomable entity and, in so doing, guide the imagination on its search for answers to profound ontological questions. This aspect of the poetics developed various tropes and patterns across seventeenth-century literature. One important and pervasive theme is the liberating – or, to borrow one of Pulter's words, 'enfranchising' – consequences of empathy with atomic being, of association with and reflection on the particles that connect 'ALL THINGS' (Traherne). Empathy is especially important to Hutchinson in *Order and Disorder*, where regenerative, compassionate attempts to restore the postlapsarian self are imagined in corpuscular terms. Related to this, poetic responses to the atom inspired self-reflection, and greater understanding of one's own capacity in recognition of the bonds between atomic, human and divine creative principles. Cavendish's atoms hold the authorial potential to 'make a World'; on the sacred side, More's immanent 'Atom-lives' require intimate self-knowledge to bring human individuals to the divine.[4] Another theme of the poetics, again related to the emphases on liberation, empathy and self-knowledge, is the powerful trope of remodelling, rebirth and resurrection. This characteristic is not exclusively theological: Cavendish's atoms form and re-form worlds made in the mind and forge the literary afterlife of their author. However, the enduring relationship between the indivisible particle and the 'Atom of Time', via the act of resurrection that takes place in the 'blink of an eye' according to St Paul, resulted in a strong association between the reconciliation of 'scattered atoms' (to quote Hutchinson) and the Christian promise of eternal life.[5] These tropes engage closely, moreover, with various philosophical positions. Within the poetry of Pulter and Hutchinson, faith in the promise of resurrection is expressed through the language of

---

[4] For Cavendish, I quote from *The description of a new world, called The Blazing World* (London, 1666), p. 97. Henry More discusses 'Atom-lives' in his 'Interpretation Generall' of *Philosophicall poems* (Cambridge, 1647), p. 422.

[5] For further information on the 'Atom of Time' resurrection, see Chapter Four, pp. 193–95.

chemical dissolution and reconstruction, which strengthens its ties with the principles of contemporary corpuscular alchemy. Another prevalent philosophy is that of vitalism, the understanding that all matter is ensouled and active. Vitalist vocabulary is closely associated with the positive poetics of the atom, and features prominently in the writings of each author discussed in this book.

These are the primary characteristics of the metaphysical 'poetics of the atom' defined and explored in this study, which came to the fore in later seventeenth-century poetical works. My priority in this book was to tease out a far-reaching, yet little-known, poetics that extended beyond the well-documented influence of Lucretius and Epicurean atomism. There were, of course, alternative sides to the seventeenth-century literary 'atom' that were connected to this legacy. In addition to this, it is important to note that creative allusions to the atom took place across a variety of literary genres, of which non-dramatic poetry was just one. Dramatic comedy and tragedy, the masque, the romance and, not least, the sermon are just some of the genres that engaged knowingly with atomic thought and in which light it would be valuable to study more closely.

It is the assertion of this book that poetry occupied a privileged place in the development of cultural understandings of the atom. Stemming in part from a Neoplatonic understanding of its spiritual power, poetry was recognised by the authors mentioned in this study, from More to Hutchinson, as the form best suited to meditation on deeper or higher subjects, including concepts of physical and metaphysical indivisibility. Poems, like atoms, created experimental spaces and increased freedom of expression. The self-conscious poetic forms of this study mimicked the very atomic acts of their focus by their efforts to break down and then recongregate their reflections on the cosmos. Adoption of an atomic poetics therefore expanded the possibilities of creative potential alongside the discoveries of divine promise. It encouraged reassessment of the place of the human individual within a world, not necessarily of instable flux and change as suggested by the reception of Lucretius, but formed from enduring indivisible particles. Neither was this a phenomenon exclusive to English culture. The suggestive bond between the atom and poetic form was recognised across vernaculars and across continents. There is room especially beyond the current study to explore the strong appeal of the 'atom' as a poetic subject for women writers from various early modern cultures: two such examples are the poetry of the Mexican nun and scholar Juana Inés de la Cruz (1648–95) and the French courtier Antoinette du Ligier de la Garde Deshoulières (1638–94).[6] It can be no coincidence that similar patterns of atomic imagery proved popular within the works of women writers from the period, especially in writings based on devotional contexts.[7]

This book formed a point of contraction, if you will, by focusing on the emergence of an 'atom poetics' in poetry from the mid to late seventeenth century. This could only ever offer a glimpse of the entwined enquiries of poetry, theology and natural philosophy that continued well into the intellectual culture of the eighteenth century. John Yolton has argued that the debates surrounding eighteenth-century materialism were anticipated in the work of the Cambridge Platonist Ralph Cudworth, a contemporary of More, who combined corpuscular physics with the presence of a higher intelligible realm in his treatise, *The True Intellectual System of the Universe*.[8] He claims that Cudworth established 'the themes, the

---

[6] See the discussion of Deshoulières's natural philosophy, including her 'Imitation of Lucretius', in John J. Conley S. J., *Women Philosophers in Neoclassical France* (Ithaca, 2002), pp. 45–74. For work on de la Cruz's natural philosophy, see Ruth Hill, *Sceptres and Sciences in the Spains: Four Humanists and the New Philosophy (ca. 1680–1740)* (Liverpool, 2000), pp. 43–94.

[7] On a related note, Jessie Hock has reflected recently on the popular reception of Lucretius amongst early modern women. She notes that 'Epicureanism had, since antiquity, been more welcoming to women than many traditional systems of thought' and continues: '[*De Rerum Natura*], as the most important text of Epicurean ethical and physical thought, not only welcomed women into its teachings, but modeled a way for women to practice natural philosophy – in poetry.' Taking this as a starting point, there is room – and need – to investigate further the striking evidence of an association between female authorship and poetic interest in atoms and atomic philosophy. See Jessie Hock, 'Fanciful Poetics and Skeptical Epistemology in Margaret Cavendish's *Poems and Fancies*', *Studies in Philology*, 115.4 (2018), 766–802 (773 and 776).

[8] John Yolton, *Thinking Matter: Materialism in Eighteenth-Century Britain* (London, 1984), pp. 4–12 (p. 5). See also Ralph Cudworth, *The true*

arguments and the counter-arguments, even much of the vocabulary' of philosophies of matter in eighteenth-century England and Scotland.[9] It is undeniable that debates surrounding matter and spirit persisted as an integral part of early modern culture, and that the 'new science' of the previous century had not yet emerged as an isolated discipline. Towards the end of the eighteenth century, the theologian and natural philosopher Joseph Priestley was still concerned with the task of combining materialism with religious truths, an enquiry that led him to deny there was any essential difference between the nature of matter and spirit.[10] Catharine Packham has shown how vitalism continued to exert a strong presence throughout the eighteenth century. She sets up her argument with the claim that the movement predates Romantic sublimity, but it also postdates the seventeenth-century vitalism of Milton, Cavendish and Marvell.[11] For Alexander Pope, Jonathan Swift, Daniel Defoe and Erasmus Darwin, amongst others, fictional literature continued not only to respond to scientific thought, but to form an integral part of natural philosophical practice.[12]

---

intellectual system of the universe: the first part; wherein all the reason and philosophy of atheism is confuted; and its impossibility demonstrated (London, 1678).

[9] Yolton, p. 12.

[10] For more information on Priestley, see Yolton, pp. 107–26.

[11] Catherine Packham, *Eighteenth-Century Vitalism: Bodies, Culture, Politics* (Basingstoke, 2012), p. 1. For a study of seventeenth-century vitalism in the works of Milton, Cavendish and Marvell, see John Rogers, *The Matter of Revolution: Science, Poetry, and Politics in the Age of Milton* (Ithaca, 1996).

[12] Sophie Gee argues that Pope's *Dunciad* 'declares an unlikely interest in theological arguments about the nature of remaindered matter' in a chapter on 'Milton's Chaos in Pope's London', from *Making Waste: Leftovers and the Eighteenth-century Imagination* (Princeton, 2010), pp. 67–90 (p. 69); Catherine Packham also discusses Pope's 'emphasis on the sheer animate powers of material nature' in *Eighteenth-Century Vitalism*, pp. 2–3 (p. 2). For more on Swift see Gregory Lynall, *Swift and Science: The Satire, Politics, and Theology of Natural Knowledge, 1690–1730* (Basingstoke, 2012). The influence of Baconian science upon Daniel Defoe is outlined in Ilse Vickers, *Defoe and the New Sciences* (Cambridge, 2006). Packham has a chapter on 'Erasmus Darwin and the Poetry and Politics of Vital Matter',

Most pertinently, references to atoms still pepper the pages of eighteenth-century literary texts. The significant relationship between corpuscular and creative thought in the period has been considered recently by Helen Thompson, in her study *Fictional Matter: Empiricism, Corpuscles, and the Novel*.[13] Thompson focuses on the literary productivity of the divisible corpuscle, which she connects to reading since both processes, corpuscular motion and the act of comprehending a text, are determined 'in relation' and acquire power 'only after the event of human perception'.[14] Thompson traces the resonances between corpuscular chemistry and eighteenth-century novel forms with great dexterity, but she is careful to distinguish between 'reductionist atomism' and the divisible corpuscles of her focus, which she claims make 'explicit the *production* of sensational *understanding* as the reader's encounter with forms and powers'.[15] According to Thompson, by the eighteenth century the indivisible 'atom' was as hard and resistant to sense – and literary productivity – as the 'blind' atoms derided by fearful preachers.[16] This assumption persists in literary criticism, especially in relation to modern texts which refer to particles that can no longer be called indivisibles. In *Science and Poetry*, Mary Midgley claimed that the triumph of the atomic philosophy resulted in the doctrine that 'competition between separate units is the ultimate law of life'.[17] According to this argument, if the early modern 'atom' inspired wonder and mysticism, contemporary literature mourns its reification. On the one hand, the particle has come to

---

pp. 147–74 (p. 147); see also Patricia Fara, *Erasmus Darwin: Sex, Science, and Serendipity* (Oxford, 2012).

[13] Helen Thompson, *Fictional Matter: Empiricism, Corpuscles, and the Novel* (Philadelphia, 2016).
[14] Ibid., p. 11; p. 7.
[15] Ibid., p. 9; p. 17.
[16] For an example of this attitude, see Edward Stillingfleet: 'to believe, I say, that all these things came only from a blind and fortuitous concourse of Atoms, is the most prodigious piece of credulity and folly, that humane nature is subject to'. Stillingfleet, *Origines sacræ, or, A rational account of the grounds of Christian faith, as to the truth and divine authority of the scriptures, and the matters therein contained* (London, 1662), p. 464.
[17] Mary Midgley, *Science and Poetry* (London, 2001), p. 3.

be associated with the isolation of modern individualism; on the other, its splitting has inspired terror and fear. So it is that the twentieth-century poet Philip Gross opens the *The Cloud Chamber* – his reflection on the cosmic significance of personal loss – with the devastating question: 'You crack an atom, what's left?'[18]

Hoban's *Riddley Walker*, however, could be placed in answer to this question to reveal another legacy to the atomic concept. Even as scientific understandings of atomism changed, something remained of the seventeenth-century poetics of the atom. No mere reductionist particle, the concept of an 'atom' still inspires poetry and wonder. So it is that Albert Camus hears of the atom in *The Myth of Sisyphus* and surmises that communication of its contents has been 'reduced to poetry'; Primo Levi concludes *The Periodic Table* by imagining the journey of a single carbon atom, which enters his brain and 'guides this hand of mine to impress on the paper this dot, here, this one'; Philip Pullman's Lyra promises Will at the end of *The Amber Spyglass* that their atoms will 'live in birds and flowers and dragonflies and pine trees and in clouds and in those little specks of light you see floating in sunbeams'.[19] For the writers considered in this book, meditation on the atomic concept led to poetic reassessments of self, society and divinity. Set loose from the binds of its scientific complexities, the infinite creative possibilities of the literary atom continue to hold great emotional power and, as Francis Bacon wrote in 'Cupid, or the Atom', to 'pierce deep' into metaphysical mysteries.[20]

---

[18] Philip Gross, 'The Cloud Chamber', in *Changes of Address: Poems 1980–1998* (Hexham, 2001), p. 65.

[19] Albert Camus, *The Myth of Sisyphus and Other Essays*, trans. Justin O'Brien (London, 1955), pp. 19–20; Primo Levi, *The Periodic Table*, trans. Raymond Rosenthal (London, 2000), p. 195; Philip Pullman, *The Amber Spyglass* (New York, 2000), p. 497.

[20] Francis Bacon, *The Wisdom of the Ancients* (English translation of *De Sapientia Veterum*), in *The Works of Francis Bacon*, ed. James Spedding, Robert Leslie Ellis and Douglas Denon Heath, 14 vols (Boston, 1857–74), XIII, pp. 67–172 (pp. 122–23).

BIBLIOGRAPHY

MANUSCRIPTS

Anonymous, Translation of Lucretius: Bodleian Library, MS Rawl. D. 314
Annesley, Arthur, The diaries of Arthur Annesley, Earl of Anglesey: British Library, Add. MS 18730 and BL Add. MS 480860
Cavendish, Margaret, 1653 copy of *Poems and Fancies*, annotated by the author: British Library, C 39. h. 27(1)
—— Letters to William Cavendish: British Library, Add. MS 70499
Evelyn, John, Translation of Lucretius: British Library, Evelyn Collection MS 34
Hutchinson, Lucy, *Order and Disorder*: Beinecke Library, Osborn MS fb 100
—— Translation of *De Rerum Natura*: British Library, Add. MS 19333
Pulter, Hester, Poems and Emblems: Leeds, Brotherton Collection, MS Lt q 32
Traherne, Thomas, 'Church's Year Book': Bodleian Library, MS Eng. th. e. 51
—— *Commentaries of Heaven*: British Library, Add. MS 63054
—— 'Dobell Folio' (including commonplace book): Bodleian Library, MS Eng. poet. c. 42
—— 'Early Notebook': Bodleian Library, MS Lat. Misc. fol. 45
—— 'Ficino Notebook': British Library, Burney MS 126
—— *The Kingdom of God*: Lambeth Palace Library, MS 1360, fols 148r–366r
—— 'Poems of Felicity': British Library, Burney MS 392

PRIMARY WORKS

Ainsworth, Henry, *The booke of Psalmes, Englished both in prose and metre with annotations, opening the words and sentences, by conference with other Scriptures* (Amsterdam, 1644)

Anon., *Rosarium Philosophorum* (Frankfurt, 1550)

Aristotle, *De Anima (On the Soul)*, trans. Hugh Lawson-Tancred (London, 1986)

Bacon, Francis, *The Oxford Francis Bacon*, ed. Graham Rees and Lisa Jardine, 15 vols (Oxford, 1996–)

—— *The two bookes of Francis Bacon. Of the proficience and aduancement of learning, diuine and humane To the King* (London, 1605)

—— *The Wisdom of the Ancients* (English translation of *De Sapientia Veterum*), in *The Works of Francis Bacon*, ed. James Spedding, Robert Leslie Ellis and Douglas Denon Heath, 14 vols (Boston, 1857–74), XIII, pp. 67–172

Bacon, Roger, *Frier Bacon his discovery of the miracles of art, nature, and magick* (London, 1659)

Bampfield, Francis, *All in one, all useful sciences and profitable arts in one book of Jehovah Aelohim* (London, 1677)

Baxter, Richard, *A saint or a brute. The certain necessity and excellency of holiness* (London, 1662)

Bentley, Richard, *The folly and unreasonableness of atheism* (London, 1699)

Beverley, Thomas, *An appeal most humble yet most earnestly by the coming of our Lord Jesus Christ, and our gathering together unto him, even adjuring the consideration of the most contrary minded who love his appearing concerning the Scripture on due compare, speaking expresly, or word for word* (London, 1691)

—— *The parable of the ten virgins in its peculiar relation to the coming and glorious kingdom of our Lord Jesus Christ opened according to the analogy of the whole parable, and of Scripture in general, and practically applied for exercising all the churches to holy watchfulness* (London, 1697)

Blackmore, Richard, *Creation: A philosophical poem demonstrating the existence and providence of a God in seven books* (London, 1712)

Blount, Thomas, *The academie of eloquence* (London, 1654)

—— *Glossographia; or, A dictionary interpreting all such hard words of whatsoever language now used in our refined English tongue* (London, 1661)

Boyle, Robert, *The sceptical chymist; or, Chymico-physical doubts & paradoxes touching the spagyrist's principles commonly call'd hypostatical* (London, 1661)

Bradstreet, Anne, *The tenth muse lately sprung up in America* (London, 1650)

Browne, Thomas, *Pseudodoxia epidemica: or, Enquiries into very many received tenents, and commonly presumed truths* (London, 1646)

—— *Religio Medici* (London, 1643)

Bibliography

Brownrig, Ralph, *Twenty five sermons* (London, 1664)
Calamy, Benjamin, *Sermons preached upon several occasions* (London, 1687)
Calvin, Jean, *A Commentary on GENESIS: Jean Calvin*, ed. John King, 2 vols (Edinburgh, 1847)
Camus, Albert, *The Myth of Sisyphus and Other Essays*, trans. Justin O'Brien (London, 1955)
Cavendish, Margaret, *CCXI sociable letters* (London, 1664)
—— *The description of a new world, called The Blazing World* (London, 1666)
—— *Margaret Cavendish's Poems and Fancies: A Digital Critical Edition*, ed. Liza Blake <http://library2.utm.utoronto.ca/poemsandfancies/>
—— *Natures pictures drawn by fancies pencil to the life* (London, 1656)
—— *Observations upon Experimental Philosophy*, ed. Eileen O' Neill (Cambridge, 1998)
—— *Orations of divers sorts, accommodated to divers places* (London, 1662)
—— *Philosophical letters: or, Modest reflections upon some opinions in natural philosophy, maintained by several famous and learned authors of this age, expressed by way of letters* (London, 1664)
—— *Poems, and fancies* (London, 1653)
—— *Poems, and phancies* (London, 1664)
—— *Poems, or Several fancies in verse, with the Animal Parliament in prose* (London, 1668)
Charleton, Walter, *The darknes of atheism dispelled by the light of nature, a physico-theologicall treatise* (London, 1652)
—— *Physiologia Epicuro-Gassendo-Charletoniana: or a fabrick of science natural, upon the hypothesis of atoms* (London, 1654)
Coke, Zachary, *The art of logick; or, The entire body of logick in English* (London, 1657)
Corye, John, *The Generous Enemies or The Ridiculous Lovers* (London, 1672)
Creech, Thomas, *Lucretius his six books of Epicurean philosophy and Manilius his five books containing a system of the ancient astronomy and astrology together with The philosophy of the Stoicks* (Oxford, 1682)
Crowne, John, *Pandion and Amphigenia; or, The history of the Coy Lady of Thessalia Adorned with Sculptures* (London, 1665)
Cudworth, Ralph, *The true intellectual system of the universe: the first part; wherein all the reason and philosophy of atheism is confuted; and its impossibility demonstrated* (London, 1678)
Cusanus, Nicholas, *Of Learned Ignorance*, trans. Fr. Germain Heron (London, 1954)

Davies of Hereford, John, *Microcosmos: The discovery of the little world, with the government thereof* (London, 1603)

Descartes, René, *The Philosophical Writings of Descartes*, trans. John Cottingham, Robert Stoothoff and Dugald Murdoch, 2 vols (Cambridge, 1985)

—— *Renati Des-Cartes Principia philosophiae* (Amsterdam, 1644)

Digby, Kenelm, *A discourse concerning the vegetation of plants* (London, 1661)

—— *Two treatises in the one of which the nature of bodies, in the other, the nature of mans soule is looked into in way of discovery of the immortality of reasonable soules* (London, 1644)

Donne, John, *The Complete English Poems*, ed. A. J. Smith, 2nd edn (London, 1996)

—— *The Poems of John Donne, Vol. 1: The Text of the Poems with Appendixes*, ed. Herbert J. C. Grierson (Oxford, 2012)

Eliot, T. S., *Selected Poems* (London, 1961)

Evelyn, John, *The Diary of John Evelyn*, ed. E. S. de Beer, 6 vols (Oxford, 1955)

—— *An essay on the first book of T. Lucretius Carus De Rerum Natura. Interpreted and made English verse* (London, 1656)

Everard, John, *The divine pymander of Hermes Mercurius Trismegistus, in XVII books* (London, 1649)

Ficino, Marsilio, *The Book 'On Obtaining Life from the Heavens'*, in *Three Books on Life: Marsilio Ficino*, ed. and trans. Carol V. Kaske and John R. Clark (Binghamton, 1989), pp. 236–405

Fuller, Thomas, *A collection of sermons* (London, 1655)

Gale, Theophilus, *The court of the gentiles, or, A discourse touching the original of human literature, both philologie and philosophie, from the Scriptures and Jewish church* (Oxford, 1670)

Glanvill, Joseph, *The vanity of dogmatizing, or, Confidence in opinions manifested in a discourse of the shortness and uncertainty of our knowledge, and its causes* (London, 1661)

Gross, Philip, 'The Cloud Chamber', in *Changes of Address: Poems 1980–1998* (Hexham, 2001)

Hale, Matthew, *A discourse of the knowledge of God, and of our selves* (London, 1688)

—— *The primitive origination of mankind, considered and examined according to the light of nature* (London, 1677)

Hall, John, *Of government and obedience as they stand directed and determined by scripture and reason* (London, 1654)

Hall, Joseph, *Select Thoughts, or Choice helps for a pious spirit* (London, 1654)

Harrington, James, *Horæ consecratæ, or, Spiritual pastime* (London, 1682)

Hermetica: The Greek Corpus Hermeticum and the Latin Asclepius in a New English Translation, trans. Brian P. Copenhaver (Cambridge, 1992)
Hesiod: Theogony, Works and Days, Testimonia, ed. and trans. Glenn W. Most (Cambridge, MA, 2006)
Hoban, Russell, *Riddley Walker* (London, 1980)
Hooke, Robert, *Micrographia* (London, 1665)
The humble advice of the Assembly of Divines, now sitting at Westminster, concerning a Confession of Faith (London, 1647)
Hutchinson, Lucy, *Lucy Hutchinson: Order and Disorder*, ed. David Norbrook (Oxford, 2001)
—— *Lucy Hutchinson's De Rerum Natura*, ed. Hugh de Quehen (London, 1996)
—— *On the Principles of the Christian Religion, Addressed to her Daughter; and Of Theology. By Mrs. Lucy Hutchinson*, ed. Julius Hutchinson (London, 1817)
—— *Order and Disorder, Or, The World Made and Undone. Being Meditations upon the Creation and the Fall* (London, 1679)
—— *The Works of Lucy Hutchinson, Vol. 1: The Translation of Lucretius*, ed. Reid Barbour, David Norbrook and Maria Zerbino, 2 vols (Oxford, 2012)
Jackson, Thomas, *A treatise containing the originall of vnbeliefe* (London, 1625)
Laertius, Diogenes, *Laertii Diogenis Vitae et sententiae eorum qui in philosophia probati fuerunt*, trans. Ambrogio Traversari (Rome, 1472)
—— *The lives, opinions, and remarkable sayings of the most famous ancient philosophers written in Greek, by Diogenes Laertius; to which are added, The lives of several other philosophers, written by Eunapius of Sardis; made English by several hands* (London, 1696)
Levi, Primo, *The Periodic Table*, trans. Raymond Rosenthal (London, 2000)
Locke, John, *An essay concerning humane understanding* (London, 1690)
Milton, John, *Milton: The Complete Shorter Poems*, ed. John Carey, rev. 2nd edn (Harlow, 2007)
—— *Paradise Lost*, ed. Alistair Fowler (Harlow, 2007)
More, Henry, *A collection of several philosophical writings* (London, 1662)
—— *Conjectura Cabbalistica* (London, 1653)
—— *Democritus Platonissans, or, An essay upon the infinity of worlds out of Platonick principles* (Cambridge, 1646)
—— *Divine dialogues, containing several disquisitions and instructions touching the attributes of God and his providence in the world* (London, 1668)
—— *The immortality of the soul, so farre forth as it is demonstrable from the knowledge of nature and the light of reason* (London, 1659)

—— *Opera philosophica* (London, 1679)
—— *Philosophicall poems* (Cambridge, 1647)
—— *A Platonick Song of the Soul: Henry More*, ed. Alexander Jacob (Lewisburg, 1998)
—— *Psychodia Platonica: or, a Platonicall Song of the Soul* (Cambridge, 1642)
Muffet, Thomas, *Theatrum Insectorum*, in Edward Topsell, *The history of four-footed beasts and serpents describing at large their true and lively figure, their several names, conditions, kinds, virtues ... countries of their breed, their love and hatred to mankind, and the wonderful work of God in their creation, preservation and destruction ... whereunto is now added, The theater of insects, or, Lesser living creatures* (London, 1658)
Newton, Isaac, *Philosophiæ naturalis principia mathematica* (London, 1687)
Ovid, *Metamorphoses: A New Translation*, ed. David Raeburn (London, 2004)
Owen, John, *Theologoumena Pantodapa* (Oxford, 1661)
Parker, William, *A revindication set forth by William Parker, in the behalfe of Dr. Drayton deceased, and himself of the possibility of a total mortification of sin in this life: and, of the saints perfect obedience to the law of God* (London, 1658)
Pettus, John, *Volatiles from the History of Adam and Eve* (London, 1674)
Phillips, Edward, *The new world of English words* (London, 1658)
Plato, *Complete Works*, ed. John M. Cooper (Indianapolis, 1997)
Plotinus, *The Enneads*, ed. John Dillon and trans. Stephen MacKenna (London, 1991)
Plutarch, *Plutarch's Morals. Translated from the Greek by several hands*, ed. William W. Goodwin (Cambridge, MA, 1874)
Pordage, Samuel, *Mundorum Explicatio* (London, 1661)
Power, Henry, *Experimental philosophy, in three books containing new experiments microscopical, mercurial, magnetical: with some deductions, and probable hypotheses, raised from them, in avouchment and illustration of the now famous atomical hypothesis* (London, 1664)
Pullman, Philip, *The Amber Spyglass* (New York, 2000)
Pulter, Hester, *Poems, Emblems, and the Unfortunate Florinda*, ed. Alice Eardley (Toronto, 2014)
—— *The Pulter Project*, ed. Wendy Wall and Leah Knight <http://pulter-project.northwestern.edu/>
Puttenham, George, *The Art of English Poesy*, in *Sidney's 'The Defence of Poesy' and Selected Renaissance Literary Criticism*, ed. Gavin Alexander (London, 2004), pp. 55–203
Rivers, George, *The Heroinae* (London, 1639)

Scott, John, *The Christian life from its beginning, to its consummation in glory: together with the several means and instruments of Christianity conducing thereunto: with directions for private devotion and forms of prayer fitted to the several states of Christians* (London, 1681)

Sennert, Daniel, *Thirteen books of natural philosophy* (London, 1660)

Sidney, Philip, *The Defence of Poesy*, in *The Art of English Poesy*, in *Sidney's 'The Defence of Poesy' and Selected Renaissance Literary Criticism*, ed. Gavin Alexander (London, 2004), pp. 1–54

Smith, John, *Select Discourses* (London, 1660)

Southwell, Robert, *The Complete Works of R. Southwell, S. J., with Life and Death* (London, 1876)

Spenser, Edmund, *The Faerie Queene*, ed. A. C. Hamilton, 2nd edn (Abingdon, 2007)

Stillingfleet, Edward, *Origines sacræ, or, A rational account of the grounds of Christian faith, as to the truth and divine authority of the scriptures, and the matters therein contained* (London, 1662)

Traherne, Thomas, *Christian Ethicks*, ed. Carol L. Marks (Ithaca, 1968)

—— *The Poetical Works of Thomas Traherne*, ed. Gladys Wade (London, 1932)

—— *The Poetical Works of Thomas Traherne, B.D. (1636?–1674), Now First Published from the Original Manuscripts*, ed. Bertram Dobell (London, 1903)

—— *Roman forgeries, or, A true account of false records discovering the impostures and counterfeit antiquities of the Church of Rome* (London, 1673)

—— *Thomas Traherne: Centuries, Poems, and Thanksgivings*, ed. H. M. Margoliouth, 2 vols (Oxford, 1958)

—— *The Works of Thomas Traherne*, ed. Jan Ross, 9 vols (Cambridge, 2005–)

Vane, Henry, *The retired mans meditations, or, The mysterie and power of godlines* (London, 1655)

Vaughan, Thomas, *The Man-Mouse Taken in a Trap, and tortur'd to death for gnawing the Margins of Eugenius Philalethes* (London, 1650)

—— *The Works of Thomas Vaughan*, ed. Alan Rudrum (Oxford, 1984)

Vita Alexander, in *Plutarch: Lives*, trans. Bernadotte Perrin (London, 1959)

Ward, Richard, *The Life of Henry More by Richard Ward*, ed. Sarah Hutton, Cecil Courtney, Michelle Courtney, Robert Crocker and Rupert Hall (Dordrecht, 2000)

Ward, Seth, *The case of Joram, a sermon preached before the House of Peers* (London, 1664)

Wilmot, John, *The Works of John Wilmot Earl of Rochester*, ed. Harold Love (Oxford, 1999)

SECONDARY WORKS

Abraham, Lyndy, *A Dictionary of Alchemical Imagery* (Cambridge, 1998)
Adams, Alison, and Stanton J. Linden, *Emblems and Alchemy* (Glasgow, 1998)
Aït-Touati, Frédérique, *Fictions of the Cosmos: Science and Literature in the Seventeenth Century*, trans. Susan Emanuel (Chicago, 2011)
Akkerman, Nadine, and Marguérite Corporaal, 'Mad Science Beyond Flattery: The Correspondence of Margaret Cavendish and Constantijn Huygens', in *Ashgate Critical Essays on Women Writers in England 1550–1700: Vol. 7, Margaret Cavendish*, ed. Sara H. Mendelsohn (Farnham, 2009), pp. 263–304
Almond, Philip, *Adam and Eve in Seventeenth-Century Thought* (Cambridge, 1999)
Archer, Jayne, 'A "Perfect Circle"? Alchemy in the Poetry of Hester Pulter', *Literature Compass*, 2.1 (2005), 1–14
'Arthur Annesley's Copy of Areopagitica', *Milton Quarterly*, 9.4 (1975), 128
Balakier, James J., 'Felicitous Perception as the "Organizing Form" in Thomas Traherne's Dobell Poems and Centuries', *XVII–XVIII: Revue de la Société d'études anglo-américaines des XVIIe et XVIIIe siècles*, 26 (1988), 53–68
—— *Thomas Traherne and the Felicities of the Mind* (Amherst, 2010)
Ball, Bryan, *The Soul Sleepers: Christian Mortalism from Wycliffe to Priestley* (Cambridge, 2008)
Barber, Benjamin J., 'Syncretism and Idiosyncrasy: The Notion of Act in Thomas Traherne's Contemplative Practice', *Literature and Theology*, 28.1 (2014), 16–28
Barbour, Reid, 'Between Atoms and the Spirit: Lucy Hutchinson's Translation of Lucretius', *Renaissance Papers* (1994), 1–16
—— *English Epicures and Stoics: Ancient Legacies in Early Stuart Culture* (Amherst, 1998)
—— 'Lucy Hutchinson and the Atheist Dog', in *Women, Science and Medicine 1500–1700*, ed. L. Hunter and S. Hutton (Thrupp, 1997), pp. 122–37
Battigelli, Anna, *Margaret Cavendish and the Exiles of the Mind* (Lexington, 1998)
Beachcroft, T. O., 'Quarles – and the Emblem Habit', *Dublin Review*, 188 (1931), 80–96
Bennett, Jim, and Scott Mandelbrote, *The Garden, the Ark, the Tower, the Temple: Biblical Metaphors of Knowledge in Early Modern Europe* (Oxford, 1998)
Blake, Liza, 'Pounced Corrections in Oxford Copies of Cavendish's Philosophical and Physical Opinions; or, Margaret Cavendish's Glitter

Pen', *New College Notes*, 10 (2018), no. 6, <https://www.new.ox.ac.uk/node/1804>

—— 'Reading Poems (and Fancies): An Introduction to Margaret Cavendish's *Poems and Fancies*', <http://library2.utm.utoronto.ca/poemsandfancies/introduction-to-cavendishs-poems-and-fancies/>

—— 'Textual and Editorial Introduction', in *Margaret Cavendish's Poems and Fancies*, <http://library2.utm.utoronto.ca/poemsandfancies/textual-and-editorial-introduction/>

Blake, Liza, and Kathryn Vomero Santos, 'Introduction', in *Arthur Golding's 'A Moral Fabletalk' and Other Renaissance Fable Translations*, ed. Liza Blake and Kathryn Vomero Santos (Cambridge, 2017), pp. 1–47

Boas Hall, Marie, 'The Establishment of the Mechanical Philosophy', *Osiris*, 10 (1952), 412–42

Borris, Kenneth, *Visionary Spenser and the Poetics of Early Modern Platonism* (Oxford, 2017)

Boswell, Jackson Campbell, *Milton's Library: A Catalogue of the Remains of John Milton's Library and an Annotated Reconstruction of Milton's Library and Ancillary Readings* (New York, 1975)

Boyle, Deborah, 'Margaret Cavendish's Nonfeminist Natural Philosophy', *Configurations*, 12.2 (2004), 195–227

Brady, Andrea, 'The Physics of Melting in Early Modern Love Poetry', *Ceræ: An Australasian Journal of Medieval and Early Modern Studies*, 1 (2014), 22–52

Brown, Stuart, 'Leibniz and More's Cabbalistic Circle', in *Henry More: Tercentenary Studies*, ed. Sarah Hutton (Dordrecht, 1990), pp. 77–96

Burns, Norman T., *Christian Mortalism from Tyndale to Milton* (Harvard, 1972)

Butterfield, Herbert, *The Origins of Modern Science, 1300–1800*, 2nd edn (New York, 1965)

Camus, Albert, *The Myth of Sisyphus and Other Essays*, trans. Justin O'Brien (London, 1955)

Chalmers, Hero, '"Flattering Division": Margaret Cavendish's Poetics of Variety', in *Authorial Conquests: Essays on Genre in the Writings of Margaret Cavendish*, ed. Line Cottegnies and Nancy Weitz (London, 2003), pp. 123–44

Chardin, Teilhard de, *The Phenomenon of Man*, trans. Bernard Wall (New York, 1959)

Clark, Stuart, *Vanities of the Eye: Vision in Early Modern European Culture* (Oxford, 2007)

Clements, A. L., *The Mystical Poetry of Thomas Traherne* (Cambridge, MA, 1969)

Clericuzio, Antonio, *Elements, Principles and Corpuscles: A Study of Atomism and Chemistry in the Seventeenth Century* (Dordrecht, 2000)

Clucas, Stephen, '"All the mistery of infinites": Mathematics and the Atomism of Thomas Harriot', in *Mathématiques et connaissance du monde réel avant Galilée*, ed. S. Rommevaux (Montreuil, 2010), pp. 113–54

—— 'The Atomism of the Cavendish Circle: A Reappraisal', *The Seventeenth Century*, 9.2 (1994), 247–73

—— 'Corpuscular Matter Theory in the Northumberland Circle', in *Late Medieval and Early Modern Corpuscular Matter Theories*, ed. Christoph Lüthy, John E. Murdoch and William R. Newman (Leiden, 2001), pp. 181–208

—— '"A Double Perception in All Creatures": Margaret Cavendish's *Philosophical Letters* and Seventeenth-Century Natural Philosophy', in *God and Nature in the Thought of Margaret Cavendish*, ed. Brandie R. Siegfried and Lisa T. Sarasohn (London, 2014), pp. 121–39

—— 'Poetic Atomism in Seventeenth-Century England: Henry More, Thomas Traherne and the Scientific Imagination', *Renaissance Studies*, 5.3 (1991), 327–40

Conley, S. J., *Women Philosophers in Neoclassical France* (Ithaca, NY, 2002)

Crawford, Julie, 'Transubstantial Bodies in *Paradise Lost* and *Order and Disorder*', *Journal for Early Modern Cultural Studies*, 19.4 (2019), 75–94

Daly, Peter, *Literature in the Light of the Emblem: Structural Parallels between the Emblem and Literature in the Sixteenth and Seventeenth Centuries* (Toronto, 1979)

Des Chene, Dennis, 'Wine and Water: Honoré Fabri on Mixtures', in *Late Medieval and Early Modern Corpuscular Matter Theories*, ed. Christoph Lüthy, John E. Murdoch and William R. Newman (Leiden, 2001), pp. 363–79

Detlefsen, Karen, 'Atomism, Monism, and Causation in the Natural Philosophy of Margaret Cavendish', *Oxford Studies in Early Modern Philosophy*, 3 (2006), 199–240

Dijksterhuis, E. J., *The Mechanization of the World Picture: Pythagoras to Newton*, trans. C. Dikshoorn (Princeton, 1986)

Dodd, Elizabeth S., *Boundless Innocence in Traherne's Poetic Theology* (Farnham, 2015)

Dolven, Jeff, 'Spenser's Metrics', in *The Oxford Handbook of Edmund Spenser*, ed. Richard A. McCabe (Oxford, 2010), pp. 385–402

Donovan, Kevin Joseph, and Thomas Festa, eds, *Milton, Materialism and Embodiment: One First Matter All* (Pittsburgh, 2017)

Dunn, Rachel, 'Breaking a Tradition: Hester Pulter and the English Emblem Book', *The Seventeenth Century*, 30.1 (2015), 55–73

Eardley, Alice, 'Hester Pulter's "Indivisibles" and the Challenges of Annotating Early Modern Women's Poetry', *Studies in English Literature 1500–1900*, 52.1 (2012), 117–41

—— '"I haue not time to point yr booke ... which I desire you yourselfe to doe": Editing the Form of Early Modern Manuscript Verse', in *The Work of Form: Poetics and Materiality in Early Modern Culture*, ed. Ben Burton and Elizabeth Scott-Baumann (Oxford, 2014), pp. 162–78

Eggert, Katherine, *Disknowledge: Literature, Alchemy and the End of Humanism in Renaissance England* (Philadelphia, 2015)

Fallon, Stephen M., *Milton among the Philosophers: Poetry and Materialism in Seventeenth Century England* (Ithaca, NY, 1991)

Fara, Patricia, *Erasmus Darwin: Sex, Science, and Serendipity* (Oxford, 2012)

Feingold, Mordechai, 'The Humanities', in *The History of the University of Oxford, Volume IV: Seventeenth-Century Oxford*, ed. Nicholas Tyacke (Oxford, 1997), pp. 211–359

Fisher, Saul, *Pierre Gassendi's Philosophy and Science: Atomism for Empiricists* (Leiden, 2005)

Fitzmaurice, James, 'Margaret Cavendish on Her Own Writing: Evidence from Revision and Handmade Correction', *Papers of the Bibliographical Society of America*, 85 (1991), 297–397

Fletcher, Harris Francis, *The Intellectual Development of John Milton: Volume II, The Cambridge University Period, 1625–32* (Urbana, 1961)

Fletcher, Puck, 'Space and Knowledge in Milton and Newton', 1–23 (forthcoming)

Fordham, Finn, 'Motions of Writing in the Commentaries of Heaven: The "Volatilitie" of "Atoms" and "ÆTYMS"', in *Re-reading Thomas Traherne: A Collection of New Essays*, ed. Jacob Blevins (Tempe, 2007), pp. 115–34

Friedlander, Paul, 'Pattern of Sound and Atomistic Theory in Lucretius', *The American Journal of Philology*, 62.1 (1941), 16–34

Gabbey, Alan, 'Henry More and the Limits of Mechanism', in *Henry More (1614–1687): Tercentenary Studies*, ed. Sarah Hutton (Dordrecht, 1990), pp. 19–35

Gee, Sophie, *Making Waste: Leftovers and the Eighteenth-century Imagination* (Princeton, 2010)

Giglioni, Guido, 'The Metaphysics of Henry More by Jasper Reid (review)', *Journal of the History of Philosophy*, 54.3 (2016), 502–03

Goldberg, Jonathan, 'Lucy Hutchinson Writing Matter', *ELH*, 73.1 (2006), 275–301

—— *The Seeds of Things: Theorizing Sexuality and Materiality in Renaissance Representations* (New York, 2009)

Gorman, Cassandra, 'Lucy Hutchinson, Lucretius and Soteriological Materialism', *The Seventeenth Century*, 28.3 (2013), 293–309

—— 'Poetry and Atomism in the Civil War and Restoration', *Literature Compass*, 13.9 (2016), 560–71

Grant, Douglas, *Margaret the First: A Biography of Margaret Cavendish Duchess of Newcastle 1623-1673* (Toronto, 1957)

Greenblatt, Stephen, *The Swerve: How the Renaissance Began* (London, 2011)

Gribben, Crawford, 'John Owen, Lucy Hutchinson and the Experience of Defeat', *The Seventeenth Century*, 30.2 (2015), 179–90

Guillory, John, *Poetic Authority: Spencer, Milton and Literary History* (New York, 1983)

Hall, Louisa, 'Hester Pulter's Brave New Worlds', in *Immortality and the Body in the Age of Milton*, ed. John Rumrich and Stephen M. Fallon (Cambridge, 2018), pp. 171–86

Hall, Rupert A., *Henry More: Magic, Religion and Experiment* (Oxford, 1990)

—— *The Scientific Revolution, 1500-1800: The Formation of the Modern Scientific Attitude* (Boston, 1954)

Hammond, Paul, 'Dryden, Milton, Lucretius', *The Seventeenth Century*, 16.1 (2001), 158–76

Hardie, Philip, 'The Presence of Lucretius in *Paradise Lost*', *Milton Quarterly*, 29.1 (1995), 13–24

Harrison, Peter, *The Fall of Man and the Foundations of Science* (Cambridge, 2007)

Hartmann, Anna-Maria, *English Mythography in its European Context, 1500-1650* (Oxford, 2018)

Hedley, Douglas, and Sarah Hutton, eds, *Platonism at the Origins of Modernity: Studies of Platonism and Early Modern Philosophy* (Dordrecht, 2008)

Henry, John, 'Atomism and Eschatology: Catholicism and Natural Philosophy in the Interregnum', *The British Journal for the History of Science*, 15.3 (1982), 211–39

—— 'A Cambridge Platonist's Materialism: Henry More and the Concept of Soul', *Journal of the Warburg and Courtauld Institutes*, 49 (1986), 172–95

—— 'Henry More versus Robert Boyle: The Spirit of Nature and the Nature of Providence', in *Henry More: Tercentenary Studies*, ed. Sarah Hutton (Dordrecht, 1990), pp. 55–76

Hill, Ruth, *Sceptres and Sciences in the Spains: Four Humanists and the New Philosophy (ca. 1680-1740)* (Liverpool, 2000)

Hine, Reginald, *Relics of an Uncommon Attorney* (London, 1951)

Hirsch, David A. Hedrich, 'Donne's Atomies and Anatomies: Deconstructed Bodies and the Resurrection of Atomic Theory', *Studies in English Literature 1500-1900*, 31.1, 69–94

Hock, Jessie, *The Erotics of Materialism: Lucretius and Early Modern Poetics* (Philadelphia, 2020)

—— 'Fanciful Poetics and Skeptical Epistemology in Margaret Cavendish's *Poems and Fancies*', *Studies in Philology*, 115.4 (2018), 766–802

Hoyles, John, *The Waning of the Renaissance 1640–1740: Studies in the Thought and Poetry of Henry More, John Norris and Isaac Watts* (The Hague, 1971)

Hutton, Sarah, ed., *Henry More (1614–1687) Tercentenary Studies: With a Biography and Bibliography by Robert Crocker* (Dordrecht, 1989)

—— 'Hester Pulter (c. 1596–1678): A Woman Poet and the New Astronomy', *Études Épistémè*, 14 (2008), 1–19

—— 'Mede, Milton, and More: Christ's College Millenarians', in *Milton and the Ends of Time*, ed. Juliet Cummins (Cambridge, 2003), pp. 29–41

—— 'More, Henry (1614–1687)', in *Oxford Dictionary of National Biography* <https://doi.org/10.1093/ref:odnb/19181>

Hyman, Wendy Beth, *Impossible Desire and the Limits of Knowledge in Renaissance Poetry* (Oxford, 2019)

Inge, Denise, *Wanting Like a God: Desire and Freedom in Thomas Traherne* (London, 2009)

Inge, Denise, and Calum Macfarlane, 'Seeds of Eternity: A New Traherne Manuscript', *The Times Literary Supplement*, 2 June 2000

Jordan, Richard Douglas, *The Temple of Eternity: Thomas Traherne's Philosophy of Time* (Port Washington, 1972)

Joy, Lynn Sumida, *Gassendi the Atomist: Advocate of History in an Age of Science* (Cambridge, 1987)

Kargon, Robert, *Atomism in England from Hariot to Newton* (Oxford, 1966)

Kerrigan, William, 'The Heretical Milton: From Assumption to Mortalism', *English Literary Renaissance* (1975), 125–66

Kuester, Martin, *Milton's Prudent Ambiguities: Words and Signs in his Poetry and Prose* (Lanham, 2009)

Lewalski, Barbara K., *The Life of John Milton: A Critical Biography* (Oxford, 2000)

Lezra, Jacques, *Unspeakable Subjects: The Genealogy of the Event in Early Modern Europe* (Stanford, 1997)

Lobsien, Verena Olejniczak, *Transparency and Dissimulation: Configurations of Neoplatonism in Early Modern English Literature* (Berlin, 2010)

Lohr, Charles, 'Ramon Lull's Theory of the Continuous and Discrete', in *Late Medieval and Early Modern Corpuscular Matter Theories*, ed. Christoph Lüthy, John E. Murdoch and William R. Newman (Leiden, 2001), pp. 75–90

LoLordo, Antonia, *Pierre Gassendi and the Birth of Early Modern Philosophy* (Cambridge, 2006)

Long, A. A., and D. Sedley, eds and trans., *The Hellenistic Philosophers*, 2 vols (Cambridge, 1987)

Lüthy, Christoph, 'David Gorlaeus' Atomism, or: The Marriage of Protestant Metaphysics with Italian Natural Philosophy', in *Late Medieval and Early Modern Corpuscular Matter Theories*, ed. Christoph Lüthy, John E. Murdoch and William R. Newman (Leiden, 2001), pp. 245–90

—— 'The Invention of Atomist Iconography', in *The Power of Images in Early Modern Science*, ed. Wolfgang Lefèvre, Jürgen Renn and Urs Schoepflin (Dordrecht, 2003), pp. 117–38

Lüthy, Christoph, John E. Murdoch and William R. Newman, eds, *Late Medieval and Early Modern Corpuscular Matter Theories* (Leiden, 2001)

Lynall, Gregory, *Swift and Science: The Satire, Politics, and Theology of Natural Knowledge, 1690–1730* (Basingstoke, 2012)

Macdonald, Paul S., *Kenelm Digby's Two Treatises* (Gresham, 2013)

Makovský, Jan, 'Cusanus and Leibniz: Symbolic Explorations of Infinity as a Ladder', in *Nicholas of Cusa and the Making of the Early Modern World*, ed. Simon J. G. Burton, Joshua Hollman and Eric M. Parker (Leiden, 2019), pp. 450–84

Marks, Carol L., 'Thomas Traherne and Cambridge Platonism', *PMLA*, 7 (1966), 521–34

—— 'Thomas Traherne and Hermes Trismegistus', *Renaissance News*, 19.2 (1966), 118–31.

Manzo, Silvia A., 'Francis Bacon and Atomism: A Reappraisal', in *Late Medieval and Early Modern Corpuscular Matter Theories*, ed. Christoph Lüthy, John E. Murdoch and William R. Newman (Leiden, 2001), pp. 209–43

Midgley, Mary, *Science and Poetry* (London, 2001)

Moreman, Sarah E., 'Learning their Language: Cavendish's Construction of an Empowering Vitalistic Atomism', *Explorations in Renaissance Culture*, 23 (1997), 129–44

Nate, Richard, '"Plain and Vulgarly Express'd": Margaret Cavendish and the Discourse of the New Science', *Rhetorica: A Journal of the History of Rhetoric*, 19.4 (2001), 403–17

Newey, Edmund, '"God Made Man Greater When He Made Him Less": Traherne's Iconic Child', *Literature and Theology*, 24.3 (2010), 227–41

Newman, William R., *Atoms and Alchemy: Chymistry and the Experimental Origins of the Scientific Revolution* (Chicago, 2006)

—— 'Boyle's Debt to Corpuscular Alchemy', in *Robert Boyle Reconsidered*, ed. Michael Hunter (Cambridge, 1994), pp. 107–18

—— *Newton the Alchemist: Science, Enigma, and the Quest for Nature's 'Secret Fire'* (Princeton, 2019)

Nicolson, Marjorie Hope, *The Breaking of the Circle: Studies in the Effect of the 'New Science' upon Seventeenth Century Poetry* (Evanston, 1950)

Norbrook, David, 'Introduction', in *Lucretius and the Early Modern*, ed. David Norbrook, Stephen Harrison and Philip Hardie (Oxford, 2016), pp. 1–27
—— 'John Milton, Lucy Hutchinson, and the Republican Biblical Epic', in *Milton and the Grounds of Contention*, ed. Mark R. Kelley, Michael Lieb and John T. Shawcross (Pittsburgh, 2003), pp. 37–63
—— 'Lucy Hutchinson and *Order and Disorder*: The Manuscript Evidence', *English Manuscript Studies 1100–1700*, 9 (2000), 257–92
—— 'Milton, Lucy Hutchinson, and the Lucretian Sublime', in *The Art of the Sublime*, ed. Nigel Llewellyn and Christine Riding, Tate Research Publication, January 2013, <https://www.tate.org.uk/art/research-publications/the-sublime/david-norbrook-milton-lucy-hutchinson-and-the-lucretian-sublime-r1138669>
Opsomer, Jan, 'In Defence of Geometric Atomism', in *Neoplatonism and the Philosophy of Nature*, ed. James Wilberding and Christoph Horn (Oxford, 2012), pp. 147–73
Ostler, Margaret J., 'How Mechanical was the Mechanical Philosophy? Non-Epicurean Aspects of Gassendi's Philosophy of Nature', in *Late Medieval and Early Modern Corpuscular Matter Theories*, ed. Christoph Lüthy, John Murdoch and William Newman (Brill, 2001), pp. 423–40
Packham, Catherine, *Eighteenth-Century Vitalism: Bodies, Culture, Politics* (Basingstoke, 2012)
Pagel, Walter, *Paracelsus: An Introduction to Philosophical Medicine in the Era of the Renaissance* (Basel, 1982)
Palmer, Ada, *Reading Lucretius in the Renaissance* (Cambridge, MA, 2014)
Passannante, Gerard, *Catastrophizing: Materialism and the Making of Disaster* (Chicago, 2019)
—— *The Lucretian Renaissance: Philology and the Afterlife of Tradition* (Chicago, 2011)
Patterson, Annabel, *Milton's Words* (Oxford, 2009)
Patterson, Annabel, and Martin Dzelzainis, 'Marvell and the Earl of Anglesey: A Chapter in the History of Reading', *The Historical Journal*, 44.3 (2001), 703–26
Picciotto, Joanna, *Labors of Innocence in Early Modern England* (Cambridge, MA, 2010)
Pines, Shlomo, *Studies in Islamic Atomism*, trans. Michael Schwarz and ed. Tzvi Langermann (Jerusalem, 1997)
Piper, William Bowman, *The Heroic Couplet* (Cleveland, OH, 1969)
Preston, Claire, *The Poetics of Scientific Investigation in Seventeenth-Century England* (Oxford, 2015)
Pritchard, Allan, 'Traherne's *Commentaries of Heaven* (with Selections from the Manuscript)', *University of Toronto Quarterly*, 53.1 (1983), 1–35

Pyle, Andrew, *Atomism and its Critics: From Democritus to Newton* (Bristol, 1997)

Raymond, Joad, *Milton's Angels: The Early Modern Imagination* (Oxford, 2010)

Rees, Emma L., *Margaret Cavendish: Gender, Genre, Exile* (Manchester, 2003)

Reid, Jasper, *The Metaphysics of Henry More* (Dordrecht, 2012)

Reisner, Noam, *Milton and the Ineffable* (Oxford, 2009)

Rimmer, Chad Michael, *Greening the Children of God: Thomas Traherne and Nature's Role in the Ecological Formation of Children* (Eugene, 2019)

Robson, Mark, 'Pulter, Lady Hester (1595/6–1678)', in *Oxford Dictionary of National Biography*, <http://www.oxforddnb.com/view/article/68094>

Rogers, John, *The Matter of Revolution: Science, Poetry and Politics in the Age of Milton* (Ithaca, 1998)

Rose, Elliot, 'A New Traherne Manuscript', *The Times Literary Supplement*, 19 March 1982

Rzepka, Adam, 'Discourse *Ex Nihilo*: Epicurus and Lucretius in Sixteenth-Century England', in *Dynamic Reading: Studies in the Reception of Epicureanism*, ed. Brooke Holmes and W. H. Shearin (Oxford, 2012), pp. 113–32

Sarasohn, Lisa T., *The Natural Philosophy of Margaret Cavendish: Reason and Fancy during the Scientific Revolution* (Baltimore, 2010)

Schiesaro, Alessandro, 'The Palingenesis of *De Rerum Natura*', *Proceedings of the Cambridge Philological Society*, 40 (1994), 81–107

Scott-Baumann, Elizabeth, *Forms of Engagement: Women, Poetry and Culture 1640–1680* (Oxford, 2013)

—— 'Lucy Hutchinson, Gender and Poetic Form', *The Seventeenth Century*, 30.2 (2015), 265–84

Shearin, W. H., *The Language of Atoms: Performativity and Politics in Lucretius'* De Rerum Natura (Oxford, 2015)

Sicherman, Carol Marks, 'Traherne's Ficino Notebook', *Papers of the Bibliographical Society of America*, 63 (1969), 73–81

Silver, Victoria, *The Predicament of Milton's Irony* (Princeton, 2001)

Sokol, B. J., 'Margaret Cavendish's *Poems and Fancies* and Thomas Harriot's Treatise on Infinity', in *A Princely Brave Woman: Essays on Margaret Cavendish*, ed. Stephen Clucas (Farnham, 2003), pp. 156–70

Spiller, Elizabeth, *Science, Reading, and Renaissance Literature: The Art of Making Knowledge, 1580–1670* (Cambridge, 2004)

Staudenbaur, Craig, 'Galileo, Ficino, and Henry More's *Psychathanasia*', *Journal of the History of Ideas*, 29.4 (1968), 565–78

Stevenson, Jay, 'The Mechanist-Vitalist Soul of Margaret Cavendish', *Studies in English Literature 1500–1900*, 36 (1996), 527–43

Stewart, Susan, *Poetry and the Fate of the Senses* (Chicago, 2002)
Sugg, Richard, *The Smoke of the Soul: Medicine, Physiology and Religion* (Basingstoke, 2013)
Sugimura, Noël, *'Matter of Glorious Trial': Spiritual and Material Substance in* Paradise Lost (New Haven, 2009)
Thompson, Helen, *Fictional Matter: Empiricism, Corpuscles, and the Novel* (Philadelphia, 2016)
Van Melsen, Andrew G., *From Atomos to Atom: The History of the Concept Atom* (New York, 1960)
Vickers, Ilse, *Defoe and the New Sciences* (Cambridge, 2006)
Walker, D. P., 'Medical Spirits in Philosophy and Theology from Ficino to Newton', in *Arts du Spectacle et Histoire des idées* (Tours, 1984), pp. 287–300
Watson, Robert N., *Back to Nature: The Green and the Real in the Late Renaissance* (Philadelphia, 2007)
West, David, *The Imagery and Poetry of Lucretius* (Edinburgh, 1969)
West, Robert H., *Milton and the Angels* (Athens, 1955)
Whitaker, Katie, *Mad Madge: Margaret Cavendish, Duchess of Newcastle, Royalist, Writer and Romantic* (London, 2003)
Wilson, Catherine, *Epicureanism at the Origins of Modernity* (Oxford, 2008)
Yolton, John, *Thinking Matter: Materialism in Eighteenth-Century Britain* (London, 1984)

INDEX

Accommodation (linguistic and poetic)   12, 20, 34, 43, 51, 67, 78, 87–88, 97–98, 115, 130–32, 146–47, 173, 180, 193–95
Adam   35, 57, 93 n.41, 175–213, 215
  and the atom   175–77, 182, 187–88, 190, 192–95, 197, 207, 210, 212, 215–16
  first principle of mankind   35, 175, 187–90, 205, 212
  four Adams   192–93, 195, 212
  innocence of   93 n.41, 175
  perfect knowledge   175–77, 182, 190–92
  suffering of   179, 182, 184–86, 206–08
  superhuman qualities   190–92
  'the Adam'   189, 193, 197, 209–10
Affective piety   103
Ainsworth, Henry   146–47
Air   46, 51, 59, 78, 105, 132, 134, 137, 169, 185, 191, 198, 206
Alchemy   13, 26, 27 n.68, 35–36, 113, 120–21, 129–32, 134, 138, 140, 145, 186–87, 198–207, 210–11, 218, 221
  alchemical bath   201–04, 210
  and emblem   129–32, 204
  and resurrection   36, 121, 130, 134, 140, 145, 186–87, 198–201, 205, 210, 218
  Aristotelian alchemy   26, 27 n.68, 120–21, 134
  calcination see Calcination
  corpuscular alchemy   13, 26, 120–21, 131, 134, 145, 186–87, 198, 204–05, 210–11, 218, 221

God as alchemist   200–01, 210
  mercury see Mercury
  Paracelsian   138
  Red King and White Queen   203
  salt see Salt
  sublimation see Sublimation
  sulphur see Sulphur
Allegory   1–4, 13, 18, 51, 53–54, 60–61, 129, 164, 204
Anaxagoras   23, 25
Angels   54, 78, 80, 102–05, 175–76, 180 n.13, 192, 194 n.44, 196 n.49, 206
Anglicanism   34, 80, 136, 144
Annesley, Arthur, Earl of Anglesey   11, 180 n.12
Apsley, Allen   178–79
Aquinas, Thomas   26
Archetype, the   3, 52, 189, 195, 215
Aristotle   23–26, 34, 78, 83 n.21, 85, 87, 113–14 n.69, 120–21, 134–36
  Aristotelianism   24 n.60, 26 n.67, 27–28, 34, 42 n.15, 87, 120–21, 134–36
  De Anima   23 n.58, 85 n.26
  De Caelo   23, 134
  De Generatione et Corruptione   23, 25
  Metaphysics   23, 134
  Physics   23–25, 134
Atheism   5, 11, 49, 188 n.30, 205, 213
Atom
  and allegory   1–4, 13, 53, 164
  and Adam   35, 175–77, 182, 187–90, 192–93, 195, 197, 207, 209–10, 212, 215–16
  and contraction   5, 19, 33, 39–40, 53, 56, 79, 92–94, 111–12, 114, 119, 166, 192, 216, 219

Index

and Cupid   1–4, 222
and human experience   14,
    35–36, 39, 54, 56, 58–59,
    70–73, 76, 79–80, 88–91,
    94, 104, 106, 115, 120, 126,
    132, 142–43, 146, 163–64,
    181–82, 186, 197, 200–01,
    205, 209–13, 217–18
and fairies   14, 20, 164
and fancy   124, 157–61, 171
and insects   20, 21 n.53
and letters   7
and poems   5, 12, 16–19, 29–30,
    33–34, 36, 39, 64, 67, 72–73,
    86–87, 97, 102–03, 112, 114,
    119, 126, 128, 142, 146, 149,
    153, 157–59, 162–63, 165–66,
    169, 173, 216, 218, 222
and politics   14–15
and selfhood, or self-
    knowledge   8, 13, 19–20,
    35, 39–40, 53–54, 58, 71, 73,
    79–80, 106, 142, 144–45,
    172, 216–17, 222
and the soul   13, 29, 34, 43, 51,
    70–73, 75–76, 80, 84–87,
    91–92, 96, 102–06, 109–15,
    125, 201, 204–05, 216
and unfathomability   3, 12,
    16–17, 36, 39, 55, 91, 96,
    114, 126–27, 131, 151, 176,
    194–95, 217
association with authorial
    self   18 n46, 119–21, 126,
    142, 145, 149, 151–53,
    156–57, 159, 163, 165–66,
    169, 172–73, 217, 219 n.7
atomies   5, 13, 21, 125
atomisation   22, 71, 125, 133,
    146, 163, 182–83, 186–87
'Atom-lives'   33, 39–40, 53–54,
    56, 58–60, 65, 67, 73, 217
'Atom of Time'   193–95, 217
'atom poem'   17–18, 19 n.49,
    34–35, 102, 105–06, 108,
    110, 119, 124, 126, 140, 149,
    151–52, 156–57, 159–62,
    166, 168–69, 172–73

congregation of   2, 10, 22, 30,
    137, 142, 151, 164, 196–97,
    204–05, 210, 218
as connecting point   4, 39, 67,
    78–79, 92, 95–96, 103, 122,
    136, 209, 211–12, 216–17
definitions of   20–21, 85,
    157–59, 163, 177–78, 193,
    216
disorder   4, 19–20, 164, 183–84,
    211–12
dissolution of   5, 20–22, 35, 121,
    125, 133–35, 139–45, 149,
    162 n.81, 165, 172, 184–86,
    196 n.49, 204–05, 210–11,
    218
indivisible *see* Indivisibility
Lucretian   7–10, 69–70, 78, 112,
    163–64, 168–69, 177–79,
    181, 183–85, 210
metaphysics of   3, 5, 12, 17, 20,
    49, 80, 114, 194 n.46, 218,
    222
negative connotations of   5–6,
    13, 49, 188 n.30, 221
order   4, 17, 19–20, 164–65,
    187, 210–11
poetics of   5, 7–8, 12–16, 19,
    21–22, 36, 86, 96, 115, 173,
    187, 210, 215–19, 222
proximity to God   1, 4, 8, 14,
    17, 20, 29–31, 34–36, 40, 53,
    55, 64, 72–73, 78–80, 84–88,
    91, 94–95, 105, 109, 111,
    114, 122, 129, 131–32, 134,
    136, 144, 146, 149, 151, 182,
    186–90, 196 n.49, 210–13,
    216–17
spiritual side of   5, 12–13,
    17, 29, 34–36, 39, 51, 54,
    70, 72–73, 76, 78, 80, 86,
    88–91, 94–96, 102–03, 108,
    112, 115, 119, 121, 125,
    131, 136, 139, 141, 144–45,
    149, 186–88, 192, 194–95,
    198, 200, 209–10, 212–13,
    215–18, 221–22
volatility of *see* Volatility

Index

world-making   12-13, 19, 35-36, 117, 119, 122, 149-51, 153, 157, 164, 170-72, 177, 216-17
Atomism   3, 5-6, 13-14, 16, 20-24, 26, 30, 33, 40, 48, 50, 56, 64, 73, 82, 85, 121, 124, 131, 134-35, 151-52, 156, 160-61, 168, 177, 183, 188, 205, 221-22
   Epicurean and/or Lucretian atomism   4-9, 11, 22-24, 27-28, 35, 38, 40, 49, 67, 83, 85, 120, 198 n.53, 205, 210, 218
   Medieval atomism   24
   Mosaic atomism   28, 83
   Neoplatonic atomism   28-30, 33, 38, 40, 67-68, 73, 85, 102
   Vitalist atomism   52, 54, 72, 120, 160 n.76, 163
Averroes (Ibn Rushd)   26

Bacon, Francis   1-4, 16 n.43, 17, 29, 36, 82, 121, 130, 190-91, 200 n.57, 220 n.12, 222
   De augmentis scientiarum   82 n.19
   De Sapientia Veterum   1-3, 222 n.20
   The Advancement of Learning   16 n.43, 130, 190-91
Bacon, Roger   24, 113-14
Bampfield, Francis   189
Baxter, Richard   45 n.19, 93
Beinecke Rare Book and Manuscript Library   4 n.8, 152 n.62, 178 n.7
Bentley, Richard   49 n.23, 188 n.30
Beverley, Thomas   194-95
Bible   13, 146, 202, 205
   2 Chronicles 34. 27   203
   1 Corinthians 15. 49   192-93
   1 Corinthians 15. 51-52   148, 193-97
   Ezekiel 22. 20   202
   Genesis   133 n.35, 178-79, 187, 190-91
   Genesis 2. 19   190-91
   Geneva Bible   13, 187, 197, 198 n.54, 202-03
   Isaiah 26. 19   199
   Job 3. 17-19   198
   King James Bible   146, 202-03
   2 Kings 22. 19   203
   Matthew 13. 31-32   94 n.45
   Matthew 25. 13   195
   Psalm 119. 28   203
   Psalm 139   146-47
   Revelations 1. 5   201
   Vulgate   193
Blackmore, Richard   212
Blood   87, 95, 102, 201-04
Blount, Thomas   20, 145-46, 194
Bodleian Library   5, n.10, 10, 18 n.48, 79 n.8, 82 n.19, 98 n.54
Book of Common Prayer   137
Boyle, Robert   9, 26, 45 n.19, 82-83
Browne, Thomas   63, 130 n.29
British Library   10 n.26, 10 n.27, 18 n.48, 77 n.3, 82 n.19, 99 n.58, 168, 170 n.95
Bruno, Giordano   24 n.60, 29, 50, 85

Calamy, Benjamin   196 n.49
Calanus   145
Calcination   133, 137, 199-200
Calvin, Jean   99
Calvinism   80 n.11, 151, 179
Cambridge Platonism   4, 33, 37, 79, 87, 110 n.67, 219
Cambridge, University of   51, 52 n.28, 118 n.2, 178
Camus, Albert   222
Cartesian philosophy   31-32, 41-45, 64
Casaubon, Isaac   110 n.67
Cavendish, Charles   121, 161, 163 n.83

243

Index

Cavendish, Margaret, Duchess of Newcastle 12 n.33, 13, 17–19, 22, 32, 34–35, 117–27, 135, 147 n.58, 149–73, 184 n.26, 216–17, 220
  'All Things Last or Dissolve According to the Composure of Atoms' 165
  atom poems 17–18, 19 n.49, 34–35, 119, 124, 126, 151–52, 156–57, 159–62, 166, 168–69, 172–73
  attitude towards authorship 18 n.46, 118–21, 126, 151–54, 161, 163, 165–66, 169, 172, 217
  'A World Made by Atoms' 117, 119, 149–52, 172
  Blazing World 118 n.2, 123, 169–71, 217
  criticism of experimental science 150, 154–55, 160
  editorial corrections 151, 154, 158, 166–68
  fancy 123, 127, 152–54, 156–62, 167, 169–71
  fiction 123, 154–55, 160 n.76, 163, 169, 171
  'It is Hard to Believe' 160, 161 n.80
  Natures Pictures 123, 153, 159
  Observations Upon Experimental Philosophy 123 n.11, 150, 156
  'Of Loose Atoms' 157
  'Of Many Worlds in this World' 159–60
  'Of Sharp Atoms' 158–59, 184 n.26
  Orations upon Divers Subjects 123
  Philosophical Letters 120 n.6, 123 n.11, 155
  Playes: Never Before Printed 167
  Poems and Fancies (1653) 12 n.33, 19 n.49, 34, 117, 126–27, 135, 150–54, 156, 158–59, 162–63, 167–69
  Poems and Phancies (1664) 118 n.1, 151, 158, 167–69
  Poems, or, Several fancies in verse, with the Animal Parliament in prose 118 n.1, 151, 158, 167
  Prefatory materials 122–23, 126–27, 153–54, 159, 161–63, 169, 172
  Sociable Letters 123
  'The Joining of Several Figured Atoms Make Other Figures' 165–66
  The Life of the Thrice Noble, High and Puissant Prince William Cavendishe 167
  Theory of perception 122, 155–57, 171
  Vitalism 120, 122, 147 n.58, 155, 160 n.76, 163, 166, 220
  'What Atoms make Change' 162 n.81, 164
  'What Atoms make Life' 158
Civil War 124
Charleton, Walter 31, 82, 84–85, 151, 189–90
Chatton, Walter 24
Chaos 1–2, 4, 21, 89 n.35, 91, 112, 129, 132–33, 164, 175, 184, 188
Chardin, Teilhard de 212
Chemistry *see* Alchemy
Christ 103, 196 n.49, 201–05
Christ's College, Cambridge 51–52, 57 n.32
Chymistry *see* Alchemy
Coke, Zachary 88–89, n.33
Cosmos 4, 18, 28, 30, 48–49, 51, 54, 60, 64, 67, 73, 87, 95, 108–09, 110 n.67, 138, 142, 169 n.94, 191, 207, 218
Conway, Anne 50
Copernicus 40, 82
Corye, John 89 n.33
Creech, Thomas 10
Crowne, John 194 n.44
Cruz, Juana Inés de la 219

Index

Cudworth, Ralph 28, 219–20
Cupid 1–4, 222
Cusa, Nicholas of (Cusanus) 24 n.60, 29, 86, 91, 109, 192

Dance 52, 149–50, 152, 164–65, 215 n.2, 216
Darwin, Erasmus 220
Davies, John, of Hereford 66
Decay 35, 59, 132, 184, 186
Defoe, Daniel 220
Democritus 22 n.55, 23–24, 28, 40, 49, 67–68, 72, 83 n.20, 85 n.26
Descartes, René 9, 29–32, 41–42, 44, 64, 70, 82, 120
  Discourse on Method 32
  Principles of Philosophy 31–32, 41
Deshoulières, Antoinette du Ligier de la Garde 219
Didacticism 6, 18, 61
Digby, Kenelm 42 n.15, 82, 121, 135–36, 142
Dissent 14–15, 180 n.12, 194
Dobell, Bertram 34, 81
Doomsday *see* Last Judgement
Donne, John 5 n.11, 13, 56, 125, 144
Dryden, John 10
Dualism 41, 180–81
Dust 20, 35, 64, 88, 111, 129–30, 132–34, 136–39, 140 n.45, 143, 145, 148–49, 182, 193, 196–97, 199–201, 207–09

Earth (element) 51, 58–59, 73, 78, 100, 102–03, 105–07, 132–33, 136–37, 148–49, 169, 192–93, 195, 197, 199–200, 208, 209 n.67
Eden 177, 190–91, 197, 206, 208
Eliot, T. S., 70
Emanation 40–41, 52, 54, 61, 73, 87
Emblem 18, 40, 50–51, 53–54, 60, 72–73, 78–80, 88–89, 110, 127–36, 139, 141, 143–46, 165

Empathy 35–36, 88, 92, 103, 114, 142, 181, 186–87, 205, 207, 210–11, 217
Embryo 143, 145–49, 172, 175 n.1, 177
Empedocles 134
  Empedoclean elements 132, 134–35, 137
Epic 19, 35, 183, 191
Epicureanism 4–11, 14–15, 24–25, 27, 30, 33, 35, 38, 40, 49, 67, 69–70, 83–85, 120, 169, 181, 186, 189, 205, 207–08, 213, 218, 219 n.7
Epicurus 10, 14, 21 n.53, 22 n.55, 23, 27, 30, 40, 67, 83 n.20, 84 n.24, 120
Eve 179 n.11, 184, 192, 196, 205–09
Evelyn, John 10, 196
Experiment 17–18, 26, 34, 38, 51–52, 68, 72, 119, 122, 125, 127–31, 138, 141, 150, 154–55, 160, 166, 175, 190, 198, 201, 216, 218
Extension 25, 41, 43–44, 47, 50, 93

Fairies 14, 20, 21 n.53, 164
Fall, the 185, 205, 207, 211, 215
Ficino, Marsilio 82, 95
Fire 51–52, 78, 85 n.26, 103, 105, 129, 132, 134, 136–38, 169, 191, 201–03, 206
First principles 4, 35–36, 50, 76, 114, 129, 132, 140 n.45, 175, 186–90, 193, 199, 204–05, 210, 212, 216
Flesh 25, 89 n.33, 102, 184, 193, 197
Flowers 57, 124, 129–30, 132, 148, 208, 222
Fludd, Robert 129
Folger Shakespeare Library 34
Fountain (Neoplatonic image) 105, 109
Fuller, Thomas 194 n.44
Furor (poetic) 16 n.43, 60–61

245

Index

Gale, Theophilus   5 n.10, 23 n.58, 28, 79 n.8, 82–84
Gassendi, Pierre   9–10, 26 n.67, 30–31, 82, 84–85, 120, 147 n.58, 151, 163
Geber *see* Pseudo-Geber
Generation and corruption   10, 119, 132, 175, 187, 195
Genesis *see* Bible
Geneva Bible *see* Bible
Glanvill, Joseph   192
Gold   102–03, 204
Gross, Philip   222

Hale, Matthew   94 n.44, 188 n.30
Hall, John   94 n.44
Hall, Joseph   89 n.33
Hammond, Henry   80 n.11
Harclay, Henry of   24
Harrington, James   196 n.49
Harriot, Thomas   121, 163
Harvey, William   52, 87
Hellenic philosophers   22
Herbert, George   18, 81, 97, 98 n.54, 99
Hermeticism   24 n.60, 28–29, 38, 85, 110–13
Heroic couplets   99, 101, 126
Hesiod   2
Hoban, Russell   215–16, 222
Hobbes, Thomas   9, 183 n.19
Holy Spirit   52, 179, 204
Hooke, Robert   154–56
Human body   36, 39–41, 45–48, 56–57, 71–73, 78–79, 84, 88–89 n.33, 94–95, 103–04, 115, 120, 132, 136–39, 143–47, 149, 180, 182, 184–85, 188 n.30, 191–204, 210, 213
Hutchinson, Lucy   4, 7 n.17, 10–11, 13, 17, 19, 21 n.53, 22, 35–36, 69 n.51, 89 n.35, 151, 175–87, 192, 196–213, 216–18
  Adam   35, 179, 182, 184, 186–87, 192, 196–98, 205–12, 216

alchemy   13, 36, 187, 198–207, 210, 218
and the Bible   13, 196–99, 201–03, 205
connection to John Milton   179–80, 183 n.19
empathy   35–36, 181, 186–87, 203, 205–07, 211, 217
Eve   179, 184, 186, 196, 198, 205–09, 211
Lucretian influence on *Order and Disorder*   13, 35, 179, 181–87, 198, 206, 210
Order and Disorder   4, 19, 35, 179–87, 192, 196–98, 201, 203, 205–13, 217
parallels with / response to *Paradise Lost*   19, 35, 179–82, 192, 197, 206–07, 209–10, 212
predestination   186, 196, 205–06
puritanical faith   151, 179, 183 n.19, 196–97
Translation of Lucretius   7 n.17, 10–11, 21 n.53, 35, 68 n.51, 89 n.35, 151, 179, 180 n.12, 181–86, 211–12
Huyjens, Constantijn   166

Immortality   4, 18, 34, 56, 59, 64, 67, 69, 80, 94, 112, 115, 131, 136, 166, 172, 193–95, 199–202, 204, 209
Incarnation   103, 193, 195, 202
Independent Church   15, 179, 180 n.12, 183 n.19, 194
Indivisibility
  of atom   2–5, 7–8, 14–18, 20–22, 23 n.58, 26, 29–36, 39–40, 42, 49–51, 53–57, 64–65, 67, 73, 76, 78, 80, 82, 84–85, 87, 89, 92–94, 99, 102, 104, 106, 109–10, 112, 115, 119–20, 125, 127, 129, 132–34, 136, 140, 145, 149, 166, 173, 175, 178, 181,

187–88, 190, 195, 198, 200, 205, 212, 216–18, 221
of God 64, 87, 132–34, 141
of soul 13, 18, 29, 43, 59, 76, 85, 94, 104, 109–10, 115, 202, 216
word coined by Henry More 39, 42

Infinity 4, 25, 31–32, 50, 54, 65–66, 72, 77, 79, 85 n.26, 90–91, 93, 102–03, 107, 109–11, 114, 122, 129, 131–33, 151, 159–60, 163 n.83, 168, 171, 191, 222
Interregnum 124, 137, 196
Intuition 3, 17, 80, 92–94, 98, 100, 102, 109, 114–15, 142, 190, 208
Islam 22 n.55

Jackson, Thomas 97–98, 100
Judgement Day *see* Last Judgement

King James Bible *see* Bible

Laertius, Diogenes 5 n.10, 23, 27, 84 n.24
Lambeth Palace Library 35, 78 n.7, 80 n.11
Last Judgement 138, 143, 193, 198
Leibniz, Gottfried Wilhelm 29, 50, 85
Leucippus 22, 28, 83 n.20
Levi, Primo 222
Lexicography 20, 194
Locke, John 93
Lucretius 6–13, 15, 20–21, 35, 40, 67–71, 78, 89, 112, 119 n.5, 120, 151, 152 n.63, 156, 163–64, 168–69, 177–87, 198 n.53, 204, 206, 210–12, 218, 219 n.7
as influence 22, 35, 38, 67–70, 152 n.63, 156, 163–64, 168–69, 177–87, 198 n.53, 210–11, 218
chaos or swarm 21, 89

De Rerum Natura 6–11, 20, 21 n.53, 23, 35, 69 n.51, 71, 119 n.5, 152 n.63, 168, 178–79, 180 n.12, 181, 183, 185–86, 211–12, 219 n.7
poetics 7–9, 11–13, 15, 21, 38, 119 n.5, 164, 179, 185, 187, 210
early modern reception of 6–12, 23, 183 n.19, 218, 219 n.7
seed 69, 186
swerve 112
Lull, Ramon 24
Lyric 8, 17–19, 61, 81, 86, 96–97, 103, 107, 112, 115, 126–27, 133, 216

Macrocosm 30 n.77, 60, 86, 87 n.30, 217
Maier, Michael 129
Marvell, Andrew 67 n.48, 125, 220
Materialism 5–8, 11–12, 14, 16, 31, 39–40, 44, 47 n.20, 49, 51, 95 n.46, 122, 178–79, 181 n.16, 183, 198 n.53, 211, 219–20
Material spirits 94–95
Means 12, 18, 34–35, 39, 73, 91–92, 94 n.44, 95–97, 99, 109, 111, 114–15, 119, 156, 188, 192, 212
Mechanism / mechanical universe 6 n.12, 23 n.57, 26 n.67, 30 n.77, 41–42, 44, 46, 49, 70, 120, 136, 150 n.60, 160 n.76, 163
Melting 187, 196, 201–07, 211
Mercy 198, 201, 203, 206–07, 211
Mercury 129–30, 204
Metaphysics 3, 5–6, 10, 12, 17, 20, 29, 30 n.77, 32–33, 37–38, 40, 42, 45, 49, 51, 53, 56, 59, 66–67, 72, 79–80, 84–87, 114, 125, 194 n.46, 218, 222
Metempsychosis 102
Microcosm 30 n.77, 86, 87 n.30

Index

Microscope / microscopy   155, 190, 192
Milton, John   17, 19 n.50, 20, 35, 52, 57–59, 145 n.55, 175, 177–83, 191–92, 197, 206–08, 210, 212, 220
   Areopagitica   180 n.12
   Comus   177
   De Doctrina Christiana   208
   Paradise Lost see Paradise Lost
Minima naturalia   24–26, 120–21, 136
Mochus the Phoenician   28, 83
Monad   29, 40, 49–54, 73, 88, 93 n.42, 192
More, Henry   4, 12–13, 17–18, 21, 28–29, 33–34, 37–73, 79–80, 82–83, 85, 87–88, 93, 95, 102, 170, 200 n.57, 216–19
   Alexander More (father)   64, 66
   Antimonopsychia   55
   'atom-lives'   33, 39–40, 53–54, 56, 58–60, 65, 67, 73, 217
   coining of new words   33, 37 n.1, 38–39, 42, 50 n.26, 58
   Conjectura Cabbalistica   28 n.71, 83 n.23
   Democritus Platonissans   18 n.47, 33, 37 n.1, 40 n.8, 49 n.24, 58 n.33, 59, 61, 64, 67–73
   Divine Dialogues   43–44, 49–51, 55 n.29, 79, 82 n.19
   'indiscerpibility'   42–45, 54
   Opera Philosophica   17 n.46, 62
   Philosophicall Poems   4, 18 n.47, 33–34, 39–40, 42, 48 n.21, 50–51, 58, 73, 88, 217 n.4
   Psychathanasia   41 n.11, 47–48, 52, 54–55, 58–59, 61, 64, 67, 70 n.53, 88
   Psychodia Platonica   18 n.47, 33, 37 n.1, 39, 41, 50, 58 n.34, 63, 67 n.49
   Psychozoia   17 n.46, 48 n.21, 61–62, 64
   self-reflection   13, 38–40, 53–54, 58–61, 65–67, 69, 73, 79–80, 217
   The Immortality of the Soul   43, 46–48
   vehicles   45–47, 54, 56, 59, 65, 72–73, 95
   vital congruity   39, 45–49, 52, 54, 65, 72–73, 87, 95
Moses   28, 83, 110 n.67
Mortalism   143–44, 208
Motes   11, 20, 21 n.53
Mythology   1–2, 21 n.53, 110 n.67, 128, 130, 215, 222

Negative theology   77 n.4, 139
Neoplatonism   4, 13, 24 n.60, 27 n.68, 28–29, 38, 40–41, 52, 54, 61–62, 64, 66–68, 70–72, 85, 87, 105, 109, 113 n.69, 218
Newton, Isaac   26, 32
Northumberland Circle   161

Orphic song   38
Ovid   129, 130 n.30
Owen, John   179, 180 n.12

Palingenesis   130–32, 135 n.39, 199
Parker, William   192–93, 195–97
Paradox   12, 14, 39, 51, 53, 55, 75–76, 82, 86, 91, 109, 114, 119, 127, 133, 140, 142, 195, 209
Paracelsus / Paracelsianism   130, 138
Paradise *see* Eden
Paradise Lost   19, 35, 57–59, 145 n.55, 175–82, 191–92, 197, 206–10, 212
Pettus, John   199–201, 207, 209
Phancy   44, 66, 167
Phillips, Edward   20
Phoenix   132, 143, 145
Plants   7 n.17, 24, 30, 54, 56–59, 94 n.45, 95 n.47, 98, 129–31, 199
Plastic spirit   41, 46–47, 70

Index

Platonic circle 30
Plenum 49, 64, 66–67, 160 n.76
Pliny the Elder 63, 129
Plotinus 39, 47–48, 53–54, 68
Plutarch / Pseudo-Plutarch 24, 51, 145
Politics 13–14, 60, 124–25, 148, 179, 183
Pope, Alexander 220
Pordage, Samuel 191, 212
Postlapsarian condition 36, 182, 184, 186–87, 190, 205, 207, 210–11, 217
Power, Henry 21 n.53, 52, 155
Predestination 150–52, 164, 186, 196, 206
Prelapsarian condition 93 n.41, 107, 175, 176 n.2, 182, 185–86, 191, 208, 210
Priestley, Joseph 220
Pronouns 71, 145, 147, 152, 211
Pseudo-Geber 26, 131, 134
Pullman, Philip 222
Pulter, Hester 13, 17–19, 22, 34–35, 117–21, 124–50, 172–73, 198, 216–17
   children 124, 137
   dust 35, 129, 132–33, 136–39, 143, 145, 148–49
   emblems 18, 127–29
   'Emblem 40' 129–36, 139, 141
   'Emblem 44' 143, 145–46
   enfranchisement / liberation 35, 120–21, 127, 141–42, 148–49, 173, 217
   focus on dissolution 22, 35, 121, 125, 129, 133–37, 139, 141–42, 144–45, 149, 172
   influence of Aristotelian corpuscular philosophy 120–21, 134–36
   manuscript 19 n.49, 118, 124, 127
   'Poem 58' 117, 125, 141–43, 145, 148–50, 172
Pulter Project, The 117 n.1, 118 n.3, 124 n.19, 139 n.42, 140 n.43, 141 n.46
   resurrection and renewal 22, 121, 125, 129–34, 136, 139–43, 145, 149, 172, 217
   selfhood 13, 119–21, 126, 142, 144–45, 147, 172
   'The Hope' 138–41
   'The Invocation of the Elements, the Longest Night in the Year, 1655' 137–38
   'The Perfection of Patience and Knowledge' 147–49
   'The Revolution' 132–35, 137, 139
Puttenham, George 16 n.43, 60
Pythagoras 28, 40, 51, 67
Pythagoreanism 24 n.60, 27 n.68, 33, 40 n.9, 50, 61, 64, 88, 102

Quintessence 78–79

Rebirth *see* Resurrection
Republicanism 15, 179, 183
Restoration, the 10, 15, 167–68, 183
Resurrection 18, 22, 36, 121, 125, 130–34, 136, 139–43, 145, 149, 172, 186–87, 194–200, 205, 209–10, 213, 217
Rivers, George 194 n.46
Rosarium Philosophorum 204
Royalism 117, 137, 180 n.12
Royal Society 10, 31, 81, 154, 155 n.68

Salt 129–30, 138
Sanderson, Robert 80 n.11
Scaliger, Joseph 26
Scaliger, Julius Caesar 120
Scepticism 31, 42, 140, 152 n.63, 154, 156, 159, 171
Scholasticism 23–26, 27 n.67, 78, 120, 131, 167
Scientific revolution, the 26
Scott, John 196 n.49

Index

Seed 30, 69, 93–95, 97 n.52, 98, 102, 132, 148, 186, 188
Self-knowledge 54, 59, 80, 90, 122 n.11, 217
Semina rerum 30
Sennert, Daniel 26, 28, 120, 204–05
Sidney, Philip 16 n.43, 60
Sin 105, 184, 186, 192–93, 195–96, 201–02, 205, 208
Skin 102, 104, 184–85, 207 n.65
Sonnet form 138, 140
Soteriology 182, 201, 211
Soul 4, 10, 13, 18, 20, 29, 34, 39, 41, 43, 45–51, 54–55, 57–59, 62–63, 67–73, 75–77, 79–80, 84–87, 90–99, 101–15, 122 n.11, 125, 127, 136, 138, 143–44, 148–49, 185, 194, 200–07, 209–10, 216
Southwell, Robert 201
Space 2, 22, 35, 41, 49 n.24, 64, 135, 142, 176, 177 n.4
  See also Vacuum, Void
Spenser, Edmund 18, 33, 37–38, 53, 56, 60, 65–67
  The Faerie Queene 18, 33, 65 n.44, 66
  Spenserian stanza 33, 37, 56, 65–67
Stillingfleet, Edward 221 n.16
Stoicism 14
St Paul 14, 193–94, 217
Sublimation 57–59, 135–36
Sublime 62, 200, 220
Sulphur 129–30, 204
Sun 7 n.17, 11, 20, 21 n.53, 52–53, 57, 72, 87 n.30, 105–06, 114, 140 n.45, 144 n.50, 169, 185 n.27, 191, 222
  sunbeam 11, 20, 21 n.53, 105, 185 n.27, 205, 222
Swift, Jonathan 220

Telescopes 155, 192
Timaeus (Plato) 51

Traherne, Thomas 4, 13, 17–19, 21, 29, 32, 34, 55, 72, 75–115, 191, 192 n.41, 216–17
  'Abyss' 88–89, 91, 96, 109
  act 13, 34, 86, 90–92, 94–97, 99–106, 108, 111, 114
  active contemplation 90–92, 96, 109, 111–12, 115
  'All in All' 104, 110, 192 n.41
  'all things' 34, 76–77, 80, 84 n.24, 86–87, 89 n.34, 100–01, 103–04, 106–09, 115, 217
  A Sober View of Dr Twisses his Considerations 80 n.11
  Burney MS 93 n.41, 96, 98
  capacity 76, 87–89, 96, 104, 106–09, 112, 114
  Centuries of Meditation 77, 81, 99
  childhood innocence 96, 106–07, 110
  Christian Ethicks 77 n.4, 104–05
  Church's Year Book 98 n.54
  Commentaries of Heaven 17 n.45, 18 n.48, 34, 55, 72, 75–78, 80, 82–95, 97, 100–06, 108–10, 112–15, 191, 192 n.41, 216
  Commonplace book 55 n.29, 79, 82 n.19, 83, 97–98, 111
  Dobell Folio 18 n.48, 34, 84 n.24, 96, 103–04
  Early Notebook 82 n.19
  'felicitie' 34, 84, 86, 99
  'From the Author to the Critical Peruser' 98–100, 104
  Ficino Notebook 82 n.19
  and Henry More 18, 29, 34, 55, 79–80, 85, 87–88, 93, 95, 102
  hermeticism 29, 85, 110–12
  insatiability 77, 80, 99, 103, 109–10
  'My Spirit' 107–11
  Roman Forgeries 80 n.11
  Select Meditations 90

'simplicitie'   92–94, 100, 106, 109
The Kingdom of God   34, 78, 82–85, 87–88, 93, 102–03, 111–12
'The Vision'   84 n.24, 86
'Wonder'   104
Trismegistus, Hermes   28, 110 n.67, 111
Twisse, William   80 n.11
Typology   2–3, 207

Vacuum   49, 168
   See also Space, Void
Vane, Henry   190
Vaughan, Henry   81
Vaughan, Thomas   37
Virbius   129–30
Vitalism   40, 49, 51–57, 64, 68–69, 72, 87, 120–22, 155, 160 n.76, 161, 163, 166, 178, 180–81, 209 n.67, 218, 220
Void   10, 22, 30–31
   See also Space, Vacuum
Volatility   102, 105–06, 108, 110, 115, 119–20, 216
Vulgate *see* Bible

Ward, Richard   62
Ward, Seth   188 n.30
Warner, Walter   121
Water   25, 51, 63–64, 78, 101, 105, 109, 132, 137, 169, 191, 204
Westminster Confession of Faith   92
Wilmot, John, Earl of Rochester   10, 180 n.12
Womb   88, 105, 137, 146–48
Word (divine)   4, 70, 197
World soul   38, 54

Studies in Renaissance Literature

Volume 1: *The Theology of John Donne*
Jeffrey Johnson

Volume 2: *Doctrine and Devotion in Seventeenth-Century Poetry
Studies in Donne, Herbert, Crashaw and Vaughan*
R. V. Young

Volume 3: *The Song of Songs in English Renaissance Literature
Kisses of their Mouths*
Noam Flinker

Volume 4: *King James I and the Religious Culture of England*
James Doelman

Volume 5: *Neo-historicism: Studies in Renaissance Literature,
History and Politics*
edited by Robin Headlam Wells, Glenn Burgess and Rowland Wymer

Volume 6: *The Uncertain World of* Samson Agonistes
John T. Shawcross

Volume 7: *Milton and the Terms of Liberty*
edited by Graham Parry and Joad Raymond

Volume 8: *George Sandys: Travel, Colonialism and Tolerance
in the Seventeenth Century*
James Ellison

Volume 9: *Shakespeare and Machiavelli*
John Roe

Volume 10: *John Donne's Professional Lives*
edited by David Colclough

Volume 11: *Chivalry and Romance in the English Renaissance*
Alex Davis

Volume 12: *Shakespearean Tragedy as Chivalric Romance
Rethinking Macbeth, Hamlet, Othello, and King Lear*
Michael L. Hays

Volume 13: *John Donne and Conformity in Crisis
in the Late Jacobean Pulpit*
Jeanne Shami

Volume 14: *A Pleasing Sinne
Drink and Conviviality in Seventeenth-Century England*
edited by Adam Smyth

Volume 15: *John Bunyan and the Language of Conviction*
Beth Lynch

Volume 16: *The Making of Restoration Poetry*
Paul Hammond

Volume 17: *Allegory, Space and the Material World
in the Writings of Edmund Spenser*
Christopher Burlinson

Volume 18: *Self-Interpretation in* The Faerie Queene
Paul Suttie

Volume 19: *Devil Theatre: Demonic Possession and Exorcism
in English Drama, 1558–1642*
Jan Frans van Dijkhuizen

Volume 20: *The Heroines of English Pastoral Romance*
Sue P. Starke

Volume 21: *Staging Islam in England: Drama and Culture, 1640–1685*
Matthew Birchwood

Volume 22: *Early Modern Tragicomedy*
edited by Subha Mukherji and Raphael Lyne

Volume 23: *Spenser's Legal Language: Law and Poetry
in Early Modern England*
Andrew Zurcher

Volume 24: *George Gascoigne*
Gillian Austen

Volume 25: *Empire and Nation in Early English Renaissance Literature*
Stewart Mottram

Volume 26: *The English Clown Tradition from
the Middle Ages to Shakespeare*
Robert Hornback

Volume 27: *Lord Henry Howard (1540–1614): an Elizabethan Life*
D. C. Andersson

Volume 28: *Marvell's Ambivalence: Religion and the Politics of
Imagination in mid-seventeenth century England*
Takashi Yoshinaka

Volume 29: *Renaissance Historical Fiction: Sidney, Deloney, Nashe*
Alex Davis

Volume 30: *The Elizabethan Invention of Anglo-Saxon England
Laurence Nowell, William Lambarde, and the Study of Old English*
Rebecca Brackmann

Volume 31: *Pain and Compassion in Early Modern
English Literature and Culture*
Jan Frans van Dijkhuizen

Volume 32: *Wyatt Abroad: Tudor Diplomacy
and the Translation of Power*
William T. Rossiter

Volume 33: *Thomas Traherne and Seventeenth-Century Thought*
edited by Elizabeth S. Dodd and Cassandra Gorman

Volume 34: *The Poetry of Kissing in Early Modern Europe
From the Catullan Revival to Secundus, Shakespeare and
the English Cavaliers*
Alex Wong

Volume 35: *George Lauder (1603–1670): Life and Writings*
Alasdair A. MacDonald

Volume 36: *Shakespeare's Ovid and the Spectre of the Medieval*
Lindsay Ann Reid

Volume 37: *Prodigality in Early Modern Drama*
Ezra Horbury

Volume 38: Poly-Olbion: *New Perspectives*
edited by Andrew McRae and Philip Schwyzer